Dr. Barry
Carol S. K
12 John Stre
Kingston, N
Phone: 914-334-9340
Fax: 914-334-9343

ENZYMES:

The Key to Health

Volume 1

The Fundamentals

Here's what others are saying about ENZYMES: The Key to Health …

Dr. Loomis, author of *ENZYMES: The Key to Health*, unravels the mysteries of the "lost chords" of healing. While medicine has made great technological strides, the causes of degenerative disease have remained obscure despite billions of dollars spent on research. The keys, which are enzyme metabolism, protein digestion and utilization and therefore optimal digestion of the food, are easily followed through his expert tutelage. This is the beginning of a millennium that will erase the differences that exist in the healing arts, and medicine and alternative medicine will all be one.

Brice E. Vickery, D.C.
Battlement Mesa, CO

Dr. Loomis has developed an incredible, simply applied method of detection of the chemical interferences of the body, and the formulations to render the innate abilities of the body to do what it knows to do best. I am using the enzyme formulations in my office and I have for seven years. It is a wonderful addition to patient treatment and restoration of health.

Stephen Fedele, Jr., D.C.
Fresno, CA

I believe that Dr. Loomis's work is cutting edge. I now know that diagnosing digestive disorders is paramount in the treatment of any illness.

Gary Moore, M.D.
Indianapolis, IN

In 1987, I met Dr. Howard Loomis and was introduced to plant enzyme therapy. It made sense to me. It explained why only about 20% or less of the people who came to me got better. Loomis had developed objective test methods to determine enzyme and nutritional deficiencies. I was very excited. I started using Loomis's 24-Hour Urinalysis test plus his palpation test and an extensive patient history to determine what enzymes and nutrients were missing from the client. I threw away my vitamin and mineral formulas, even though some of my clients screamed at me. It worked. Clients got better faster, stayed well longer, and with fewer complaints.

Now, in 1999, I am even more excited, because the field of plant enzyme therapy is growing by leaps and bounds. There is no limit to how enzyme therapy can help remediate the many problems caused by poor digestion coupled with a poor diet.

Clearly, it would be impossible for me to help a single person who seeks my advice without the Loomis line of enzymes. These enzymes have changed my life and have made it possible for me to help hundreds of people with digestive disturbances and other serious problems caused by enzyme deficiencies.

Loomis is truly this century's foremost enzyme pioneer, and I am grateful for his genius and his research.

Lita Lee, Ph.D.
Eugene, OR

Healers often cannot teach, and teachers cannot heal. Dr. Loomis has the ability to do both. His vast knowledge of human physiology and his understanding of the body's basic need of being able to digest what it eats has enabled him to develop a system of plant-derived enzyme products that play a significant role in predigesting food and delivering nutrition past an incompetent digestive system. His work has changed my life, health, and my own practice as a healer. My patients and I will be eternally grateful.

Marianne Miller, D.C.
Anchorage, AK

I feel that I have benefited tremendously from Dr. Loomis's work and the knowledge that he has shared in a learnable and useable system. To be able to take the guesswork out of nutrition and have a way to stop constant stress to a person's system is such an important first step in helping someone to be well that I would not want to have to go back and work without it. What may be most unique about Dr. Loomis's work, with nutritional theories so abundant these days, is his work is actually about nutrition.

Rich Easterling, N.D.
Grass Lake, MI

Thank you very much for your valuable information during your five-day workshop on Enzyme Replacement Nutrition. I was able to put your enzyme therapy to work in conjunction with NAET and acupuncture right away. The knowledge Dr. Loomis has provided to the entire medical world has thrown open the door of complementary medicine in the use of enzymes and other viable nutrients.

Elizabeth Chen Christenson, M.D.
Maumee, OH

Enzyme Replacement Nutrition is nothing short of a miracle. I was going down the tubes fast with my health and seriously thinking that I would have to close my practice. After attending the Associate training seminar, I regained my health using enzymes. Now, even the most complex and difficult cases are easy to treat. Many thanks to Dr. Loomis and Enzyme Replacement Nutrition.

Heather Koeppel, D.C.
Chamblee, GA

After fifteen years as a primary care physician, I would not wish to consider treating the myriad of health problems that we regularly encounter without the expertise you impart in your seminars or without the enzyme supplements you have created.

It has been a pleasure and an adventure knowing you, working with you, and learning from you. May you continue your marvelous career as the ultimate expert on enzymes and enzyme formulations for many years to come.

Ivan F. Kelley, M.A., D.C., C.C.N., D.A.C.B.N.
Newport, OR

Dr. Loomis has found the missing link to health. I have found profound results using such a simple concept. Balance the body and get repeated outstanding results. This program has been an astounding complement to my practice.

Jean Dugan, L.Ac.
Portland, OR

ENZYMES

The Key to Health

Volume 1
The Fundamentals

by Howard F. Loomis, Jr., D.C., F.I.A.C.A.

A publication of 21st Century Nutrition Publishing
6421 Enterprise Lane, Suite 110
Madison, WI 53711
800-662-2630
Fax 608-273-8111

Designed and published by
Grote Publishing
634 West Main Street, Suite 207
Madison, WI 53703-2634
608-257-4640

Printed in the U.S.A.
10 9 8 7 6 5 4 3 2 1

Library of Congress Cataloging-in-Publication Data
Loomis. Howard F., 1937—
    Enzymes : the key to health / by Howard F. Loomis, Jr.
    v. < 1–   > cm.
    Includes bibliographical references and index.
    Contents: v. 1. Fundamentals.
    ISBN 0-9663436-2-X
    1. Enzyme--Physiological affect. 2. Enzymes--Therapeutic use.
    3. Nutrition. I. Title.
    QP601.L656 1999
    615'.35--dc21   99-17910
                                                              CIP

This book takes the position that, while sick people have increased nutritional needs, nutritional protocols should never be used to treat disease conditions. Further, any change in diet or supplement use should be done under the supervision of a licensed health care professional.

Dedication

This book is dedicated to my grandchildren, Jeremy, Courtney, Alex, and Jordan. Like all grandchildren, they are clearly the most wonderful, most beautiful, most important people ever to be born. It is my prayer that they and their contemporaries will grow and mature in a world that knows how to prevent chronic degenerative disease.

Acknowledgments

To the memory of my father who taught me that improved digestion and elimination were the key to improving the health of his patients.

To the pioneers in the field of autointoxication, Drs. Tilden and Sheldon, and especially to Dr. Edward Howell for his graciousness and for giving me the key that unlocked the door.

To Anthony Collier for giving me my start in the enzyme business, and to my patients who were so patient and understanding as the Enzyme Replacement Nutrition was tested and developed.

To Jerome Fisher whose belief in natural healing is truly boundless, and whose counsel and guidance have made it possible for me to move from private practice into the business world and bring this message to a wider audience.

To Pamela Warner who insisted this book be written. And for her ideas, support, and marketing skills, and especially for her patience in listening to my ideas. Thanks to her, only the good ones see the light of day.

To Franciska Anderson whose faith in my work and editorial skills have brought this book through the publishing stage. This book never would have progressed beyond the "wish stage" without her dedication and patience.

To my children for their years of patience as they wondered what in the world I could be working on that demanded that much time and effort. Fortunately for me, they not only see the fruit of the labor but are part of it.

Biography

Howard F. Loomis, Jr., D.C., F.I.A.C.A. is a 1968 graduate of Logan Chiropractic College. Dr. Loomis's interest in nutritional food enzymes began when he had the privilege of working with Edward Howell, M.D., the food enzyme pioneer.

He has taught this system extensively to professional health care practitioners. As president of 21st Century Nutrition, Inc.® for fifteen years, he has forged a remarkable career as an educator, having conducted over 400 seminars to date in the United States, Canada, Germany, and New Zealand on the diagnosis and treatment of enzyme deficiency syndromes.

He has an extensive background in enzymes and enzyme formulations. He is currently president of his own enzyme company, Enzyme Formulations, Inc.® With his exciting approach to health and wellness, he is now preparing others to take health care into the 21st century.

Foreword

In my medical journey, I have been exposed to many new and interesting concepts. Few of them have been as exciting as has enzyme therapy. When I first learned of this and attended Dr. Loomis's workshop, it opened up a whole new world to me. I had learned in biochemistry and physiology the importance of digestive enzymes and how enzymatic processes work. I did not, however, have a clue as to how effective they could be in clinical medicine. Being mentored by Dr. Loomis has provided me with an entirely new grasp of the subject. Not only does he share information that markedly enhances one's understanding of biochemistry, but he provides the interested clinician with a wonderful therapeutic tool to significantly enhance his clinical efficacy. In short, you will not only do the job smarter but you will do it better (more effectively).

As you read and study this information, do so with an open and creative mind. Dr. Loomis has done a wonderful job of organizing it, presenting it in a clear and understandable manner, and making it clinically applicable. By the way, you M.D.s and D.O.s who may be reading this, don't think that you learned everything in medical school that you will be exposed to here. I promise you, this information will stretch your mind! This is truly practical biochemistry—it's where the rubber meets the road. Have fun—I sure have!

Arthur Davis, Jr., M.D.
Indio, CA

TABLE OF CONTENTS

INTRODUCTION TO ENZYME REPLACEMENT NUTRITION

A DIFFERENT PERSPECTIVE

To build a house you need a design, a plan drawn up by someone who knows how. Each builder throughout history has stood on the shoulders of those builders who have gone before.

Introduction to Enzyme Replacement Nutrition

What is Enzyme Replacement Nutrition, and why is it important to you? We all know that food is composed of protein, carbohydrates (including fiber), fats, vitamins, and minerals. But until recently, few of us appreciated that all living things also contain enzymes. More to the point, life is not possible without enzymes. In fact, it is the enzymes that are responsible for the vast majority of all the biochemical reactions that bring our foods to maturity or ripeness. These enzymes will also digest the food in which they are contained when conditions are right for that to happen. For example, an apple falls from a tree, and a few days later a brown spot is seen on the apple where it landed on the ground. We refer to that spot as being "spoiled" or "bruised," but it has only been digested. When the apple landed, it broke the cell walls, and the enzymes contained in those cells were liberated to begin digesting the apple. Those same enzymes will begin digesting the apple when you chew it.

Enzymes are energy, and energy is defined in high school physics textbooks as "capacity to do work." Enzymes have the energy to perform the biochemical and physiological reactions that occur in all living things. The other components of our food supply, namely protein, carbohydrates, fats, vitamins, and minerals, are only building blocks. They do not perform work.

Over the past 50 years, the American public has become increasingly conscious of nutrition and its importance in our lives. Most of us know that our

> **KEY POINT**
>
> Enzymes are the construction workers of the body. They use the vitamins and minerals as building materials to put up the latest architectural marvel—your body. At times there is an ample supply of building material but not enough workers to complete the job. The building materials will remain unused until enough workers show up at the job site.

diets should be composed roughly of 50% carbohydrate, 30% protein, and 20% fat. Most of us know that vitamins and minerals are not made by our bodies and we must consume them to prevent deficiencies. In fact, certain vitamins and minerals are routinely added to our food supply to prevent problems such as goiter (iodine), rickets (vitamin D), and birth defects (folic acid).

But very few people know about the vital role nature planned for food enzymes to play in the digestion of our food or that those enzymes must be removed from our diets to prolong food's shelf life. I think you might agree with me when I say that digestive problems are now a major health problem in this country. Just look at the billions of dollars spent each year on antacids. I think there is a correlation here, and that is what this book is all about.

My goal in writing this book is to present an easy-to-understand explanation for considering the role naturally occurring food enzymes can play in our lives. This book will present a simple system, if you will, of understanding how our bodies work and the role that enzymes play in keeping us healthy. I will try to show you when and why your body develops symptoms, and the role diet and digestion play in their appearance. I will also try to show you why two people eating the same diet, for example, you and your spouse or other family member, may not develop the same problems. In other words, I will help you to obtain something I call *nutritional objectivity* and use it to prevent disease and maintain health.

I first got involved in enzymes around 1980, and it changed my life. I practiced as a chiropractor in Missouri from 1967 until 1993. I'm a second-generation chiropractor, and my father believed greatly in using betaine HCl and pancreatin supplements to improve digestion. He felt that digestion was the place to begin healing or to maintain health—not to cure disease but to maintain health. Since I had seen this procedure benefit so many patients, it was natural for me to use it when I started my own practice in 1967.

My interest was in understanding why, when there is no history of injury, some people develop back problems and others do not. Was there a nutritional component? It seemed logical that there would be because the ability to digest and assimilate protein and consequently improve the body's ability to carry calcium and other minerals to the tissues is very important. Most people who acquire symptoms of musculoskeletal dysfunction, such as osteoporosis, herniated discs, bursitis, and leg cramps, do not readily digest protein.

I worked for 12 years (from 1967 to 1979) correlating laboratory results from urinalysis and blood work with physical examination findings and using the traditional digestive supplements such as betaine HCl, pancreatic enzymes, and ox bile salts trying to find a way to improve digestion and make nutrition work as a science. I finally gave up in frustration in 1979, having failed to find consistent correlations. I could never find the clinical parameters that would allow me to say this person needs calcium, this one needs magnesium, that one needs calcium *and* magnesium, that one needs vitamin C, or that one needs better protein digestion. I was convinced there was no objective means of utilizing nutritional supplements.

In 1980 I was fortunate to be introduced to the work of Edward Howell, M.D., and his "food enzyme concept." After reading his two books, *Enzymes for Health and Longevity* and *Enzyme Nutrition*, I was convinced he had found the missing link for providing consistent results in clinical nutrition.

Dr. Howell had graduated from medical school at the University of Illinois in 1919, the year before the first vitamin was discovered. After graduation he practiced at the Lindlahr Institute in Chicago, which was the Mayo Clinic of his day. They specialized in the treatment of chronic degenerative diseases using a system of fasting and raw-food diets. This was in a time prior to the discovery of insulin, and diabetes was the major degenerative disease. Cancer was not as readily diagnosed as it is today, and diabetes was the number-one killer among the general population.

Dr. Howell was impressed with the results obtained with raw foods and fasting, and he struggled to find an explanation. He became convinced that there had to be something else in food besides protein, carbohydrates, fats, vitamins, and minerals. This led to his eventual fascination with enzymes found in raw food and the role they play in digestion and maintaining health. In 1945 he formulated two enzyme formulas, one using food enzymes and the other using pancreatic enzymes. He formed his own company and devoted the rest of his life to formulating his theories of the existence of a predigestive stomach in humans, and the occurrence of enlargement of the pancreas (hypertrophy). Hypertrophy occurs when an organ is required to make all the enzymes needed to digest the diet instead of being able to benefit from the digestive work performed by the enzymes found in the raw food to assist it.

Dr. Howell retired from active practice in the early 1970s. He tried to disseminate information about food enzymes and their importance, and the scientific community pretty much ignored him. His work was theoretical. He talked about the predigestive stomach, which I was not aware of until I read his books. In 1982 I was flown down to Fort Myers, Florida, by a supplement company to see Dr. Howell, and I spent time with him in his home. He allowed me access to his accumulated notes, many of which I copied, including his extensive bibliography to *Enzymes for Health and Longevity*. He was very gracious in sharing his time with me, and he completely changed my attitude about nutrition and taught me a great deal about the importance of food enzymes. Gaining permission to copy many of his accumulated notes and bibliography was incredibly valuable since they were destroyed when he died in the late 1980s.

Although he had worked with fasting and raw-food diets throughout his professional career, Dr. Howell did not have access to the concentrated plant enzyme supplements available today. When I became familiar with his theoretical work on the predigestive stomach and plant enzyme supplementation, I was invited to give my impressions on the possible clinical uses of his work and to formulate the first line of plant enzyme supplements for professional use.

My initial impression was to combine enzymes with nutrients from whole food or herbal sources so they could be assimilated past a less-than-perfect digestive system. For example, it is very difficult for a patient who has difficulty digesting fats to assimilate food supplements of concentrated oils. I thought that since nutritional supplements employed the "magic bullet" theory of removing symptoms, all I needed to know were the symptoms of enzyme deficiencies. For example, what symptoms would protease relieve? How about lipase, amylase, and others? I was surprised to learn that Dr. Howell did not know. But, on second thought, it was understandable; although he had developed the theory of predigestion, he had not worked with the individual enzymes or with the concentrated dosages that modern technology could provide. Unfortunately, since this was an entirely new concept, there simply was no place to turn for guidance.

What was needed were clinical outcome studies to determine enzyme usages and formulations that could truly nourish the body and assist in moving nutrients past incompetent digestive systems. What was needed was a method of examination that could tell what nutrients the body needed—either because they were not adequate in the diet or because the body was not digesting and assimilating what was ingested. What was not needed was a method of examining for isolated chemical compounds. Since there are only so many ways to scientifically examine a body, I quickly eliminated all the choices but one.

Case histories are based on symptoms and are not acceptable since they do not reveal the cause of the symptoms.

Blood work is not helpful because the body is required to maintain the blood within normal limits for as long as possible. In other words, the body will compensate for any deviation in the contents of the blood by pulling what it needs from its tissues. For example, in the condition known as osteoporosis, mineral levels remain normal in the blood while the body gradually removes minerals from the bones. Putting it another way, blood tests are used to find objective evidence of deviations from normal (disease). They do not serve as an early-warning signal that some problem is beginning to develop. By the time the blood tests are not normal, the problem is already there.

Obviously, the same problem exists with physical signs. By the time the physical signs of a nutrient deficiency are evident, the damage has already been done. The signs of scurvy (vitamin C deficiency), for example, take about 60 days to become obvious. But there is no early-warning signal that the body is deficient until the individual begins to experience severe fatigue and the gums begin to bleed.

Clearly, X-ray and other advanced technologies are not useful for the same reason. The damage must be done before the deficiency becomes obvious. I used to tell my patients that medical examinations could only identify the disease process going on in their bodies, and if all the tests were negative, they should rejoice because it meant they were not diseased and the solution was probably nutritional.

The only possible means of identifying nutritional needs for enzymes is to conduct urinalysis examinations. While that may surprise you, consider that the normal values for urinalysis are well established. They have been documented for well over 100 years. Examining what is being eliminated through the kidneys and bladder tells us what the body is throwing away because it has too much of it, and what it is holding because it does not have enough. The explanation is a little more complicated than that, but that gives you the general idea. Next, I found that the urine has to be collected for 24 hours, since random samples taken throughout the day vary considerably because of what is eaten and drunk! That, of course, is the entire point. By collecting the urine for a 24-hour period, we can begin to draw some interesting conclusions.

Next, I began to use individual enzyme supplements, first in myself and my family and then gradually in consenting patients. This established exactly how those enzymes influence the body. Finally, after five years of extensive clinical study, trial and error, and laboratory workups, the first products were ready to go to market.

In 1985 I began lecturing on the clinical applications of plant enzymes while continuing my practice and clinical investigations. Gradually, the work attracted attention within the healing arts. By the late 1980s Dr. Howell had died, and other companies began to copy the information and products; plant enzymes gained more and more attention. Soon, sales personnel involved with the original company left to form their own multilevel companies. It has been very gratifying to see this trend continue as the importance of plant enzymes becomes increasingly appreciated.

In 1993 I sold my practice, formed my own company, and now devote my time to formulating, writing, and teaching. The clinical studies continue but are now performed by hundreds of associate doctors around this country and the world. This program has been quite successful, and these doctors are now carrying on this work in their own private practices. These health care practitioners are taught a comprehensive method for establishing the need for enzyme supplementation that includes case history, dietary analysis, physical examination, and interpretation of laboratory tests, including blood and 24-hour urinalysis. This model has stood the test of time and is very useful in making nutritional evaluations. Many new clinical applications have been found in the last 18 years as much more has been learned about these essential nutrients. I hope the information in this book will be as helpful to you as it has been to me.

Part One

ESSENTIAL
KNOWLEDGE

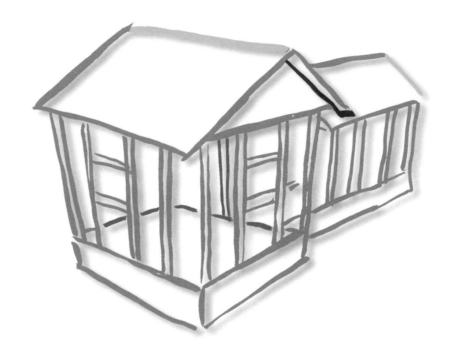

The Development of Nutrition as a Science

A Society of Cells

To erect a house you must have good building
materials. The frame and foundation will only be as
good as the material they
are made from. In
nutrition, protein,
carbohydrates, fats,
vitamins, and minerals are
building materials—
together they are called
food.

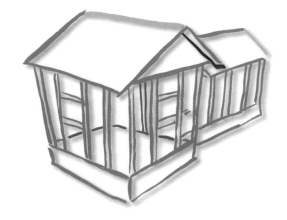

Chapter 1

The Development of Nutrition as a Science

The science of nutrition is a relative youngster in the scientific community, having been recognized as a distinct discipline only in 1934 with the organization of the American Institute of Nutrition.

Real public interest in nutrition began only after World War II, when food supplements became available. It has captured significant interest from the public and policy makers alike in the past 25 or 30 years as attempts to control and profit from the sale of these supplements has grown into a multibillion-dollar-a-year industry. Likewise, the identification and rationale for the necessity of other vital nutrients in food has increased dramatically.

Nutrition had to await advances in chemistry and biology before maturing into a distinct discipline. You see, nutrition cannot stand alone any more than any other science can. It both depends on and contributes to biochemistry, microbiology, physiology, cellular biology, medicine, and food science. One of the more recent technological breakthroughs, genetic engineering, offers both a tool and a challenge for nutritionists.

Nutrition as a science has been defined in many ways. Most simply, it has been identified as the science of nourishing the body properly or the analysis of the effect of food on the living organism. The Council on Foods and Nutrition of the American Medical Association states that nutrition is "the science of food, the nutrients and other substances therein, their action, interaction, and balance in relation to health and disease and the processes by which the organism ingests, digests, absorbs, transports, utilizes and excretes food substances" (cited in *Introductory Nutrition* by Helen A. Guthrie, 6th edition, St. Louis: Times Mirror/Mosby College of Publishing, 1986, p. 4).

There is general agreement that nutrition is concerned with the way food is produced, with any changes that occur in it before it is eaten, and with the way the body uses food until it is either built into body tissues or excreted. This includes the study of digestion, absorption, and transportation of nutrients to and from cells and the way in which nutrients are used within the many types of body cells. In addition, the public is becoming increasingly concerned with the foods that they choose to eat and how these choices affect their health and well-being. They are particularly concerned with how the food is grown, what chemicals are used, and what preservatives have been added.

Historical Background

Although most of the organized study of nutrition has been confined to the 20th century, we have evidence of a long-standing curiosity about the subject. A few well-conceived nutritional experiments were performed earlier than the 1900s, but these stimulated only vague interest. For convenience, the history of nutrition has been divided into four eras:

- The naturalistic era (400 B.C. to 1750 A.D.)
- The chemical-analytical era (1750 to 1900)
- The biological era (1900 to 1955)
- The cellular or molecular era (1955 to the present)

The Naturalistic Era (400 B.C. to 1750 A.D.)

During the naturalistic era, people had many vague ideas about food, most of which revolved around taboos, magical powers, or medicinal value. Early men and women recognized that food was essential for survival and did not discriminate about the relative value of different foods, just as millions do today. In Biblical times, however, Daniel observed that men who ate pulses (legumes such as beans and peas) and drank water thrived better than did those who ate the king's food and drank wine. Hippocrates, the father of medicine, considered food a single universal nutrient when he discussed food, health, and disease in 400 B.C. He believed that "all diseases begin in the stomach" and weight loss during starvation was caused by "insensible perspiration." It was not until the 16th century that the first concepts of physiology were discussed.

In the early 17th century the Italian physician Sanctorius, curious about the fate of food in the body, weighed himself before and after each meal. His only explanation for his failure to gain weight in keeping with the amount of food taken in was that there must be weight loss, resulting from insensible perspiration. It was during this period that such men as Harvey and Spallanzani, with their interest in circulation and digestion, made observations that eventually facilitated the study of nutrition. In 1747 the first controlled nutrition experiment was carried out by a British physician, Lind. He attempted to find a cure for scurvy by treating 12 sailors ill with the disease with six different substances. He determined that either lemon or lime juice was effective in curing scurvy; the other substances—oil of vitriol, cider, nutmeg, seawater, and vinegar—were not.

The Chemical-Analytical Era (1750 to 1900)

In the 18th century, Lavoisier, who became known as the father of nutrition, initiated the chemical-analytical era. He studied respiration, oxidation, and calorimetry, which are all concerned with the use of food energy. His work with guinea pigs was the first investigation of the relationship between heat production and oxygen use in the body.

Early in the 19th century, methods were developed for determining carbon, hydrogen, and nitrogen in organic compounds. Analyzing foods to find amounts of each of these elements led Liebig to suggest that the nutritive value of food was a function of its nitrogen content. He also postulated that an adequate diet must provide plastic foods (protein) and fuel foods (carbohydrate and fat). Dumas, a French chemist, tested this hypothesis during a siege of Paris in 1871. His efforts to produce a synthetic milk of carbohydrate, fat, and protein in the proportions believed to be found in cow's milk proved unsuccessful, and the infants to whom he fed it died. Dumas logically concluded that milk must contain some unknown nutritive substance.

A similar conclusion was reached in 1881 by Lunin. He found that mice died when fed a diet of purified casein (a protein), milk sugar (a carbohydrate), milk fat, and the inorganic ash from milk, while those that were fed milk thrived. Between 1881 and 1906 purified diets were used in 12 other animal experiments. All astonishingly led to essentially the same conclusion: the addition of small amounts of natural foods was necessary to promote growth and to maintain health in the animals studied. Food obviously contained other substances besides carbohydrate, fat, protein, and mineral ash, but their nature remained a mystery.

KEY POINT

The consequences of a poor and inadequate diet were clearly identified in the work of both Dr. Weston Price and Dr. Francis Pottenger. There is more information on their studies on pages 26–29.

The Biological Era (1900 to 1955)

By 1912 it had been well established that another dietary essential existed besides carbohydrate, fat, protein, and mineral ash. Casimir Funk, recognizing that this elusive dietary component was essential to life *(vita* in Latin) and believing it to be *amine* (or nitrogen-containing), introduced the term *vitamines* to describe it. Two independent studies showed that there were at least two "vitamines": fat-soluble "vitamine" A and water-soluble "vitamine" B. McCollum's work at the University of Wisconsin showed that some fats, such as butter, contained an essential growth factor whereas others, such as lard, did not. Eijkman observed that a water-soluble substance in rice bran prevented beriberi, a disease common in the Orient. By 1920, when it was established that all vitamins did not contain nitrogen, the final "e" was dropped to obtain the term *vitamin*, which is now a household word.

The concept that diseases such as beriberi, scurvy, rickets, and pellagra were the result of an absence of nutrients needed in very small amounts did much to stimulate attempts to identify the nature of these dietary essentials. It soon became clear that there were several components of both fat-soluble vitamin A and water-soluble vitamin B. By 1940, four fat-soluble and eight water-soluble vitamins had been identified as essential elements of the human diet; several others had been identified as essential for various species of animals. The chemical structure of each vitamin had been established, many had been synthesized, and knowledge of their biological roles was accumulating rapidly. Since 1940 only two other essential vitamins have been identified—folacin and vitamin B_{12}.

During this same period, mineral ash from the diet was studied. Minerals, like vitamins, proved to be a complex mixture of elements—20 of which have been established as dietary essentials for humans. The essentiality of several others is still uncertain to this day.

The Cellular or Molecular Era (1955 to the Present)

Since 1955, many technological developments have made it possible to study the nutrient needs of the individual cells and even the subcellular components, or organelles, of the cell. As a result, an understanding of the intricacies of cell structure and the complex and vital role that nutrients play in the growth, development, and maintenance of the cell is accumulating rapidly. Nourishment of cells is essential for the nourishment of tissues; in turn, nourishment of tissues is basic to the nourishment of organs and ultimately of the whole body. Failure to form an essential enzyme or other cellular components results in the malfunction or death of a cell. This process eventually results in a specific physical symptom of ill health.

After 1960, emphasis in nutrition research changed from a search for essential dietary components to a study of the interrelationships among nutrients, their precise biological roles, the determination of human dietary requirements, and the effect of processing on the nutrient quality of foods.

It is important to remember that, although we depend on studies at the molecular level to enhance our understanding of nutrition, food is the ultimate source of nourishment. Food is the only source of the many nutrients on which cell growth and function depend.

Present Status

It has only been a hundred years since it was shown that carbohydrates and proteins were not the only food components needed for normal growth and development. Today we have a vast, complex, and rapidly expanding knowledge of more than 45 nutrients that must be supplied in the diet. The absence of any one of these nutrients, whether it is needed in small or in large amounts, can have a profound effect on health.

Although the last vitamin was discovered more than 30 years ago, we are still identifying essential mineral elements. The number of recent scientific publications in the field of nutrition indicates how much nutrition research has increased since the concept of vitamins was first presented:

K E Y   P O I N T

The most extensive research on vitamins and other essential nutrients took place in the 1960s at the University of Texas by Roger J. Williams, Ph.D.

- In 1913, four nutrition articles were published, all by Casimir Funk.
- In 1920, the number of articles had risen to 73.
- In 1930, 724 articles were published.
- By 1985 the number of articles on single nutrients had increased substantially, with 200 to 500 references and more than 4,000 other papers with implications for nutritional science.
- A trip to the library in 1985 found there were about 2,000 articles on naturally occurring food enzymes, and all but a few were concerned with how to destroy them.

Discovering more about the complex interactions among vitamins, minerals, and macronutrients is part of a challenging frontier in science. Translating our knowledge of nutrient needs into meaningful dietary advice for a public concerned about the impact dietary practices have on health, longevity, and behavior is an even greater challenge.

The fact that scurvy, rickets, beriberi, pellagra, and kwashiorkor—all nutritional-deficiency diseases—are found in developing countries is stark evidence of our failure to apply all the nutrition information that we have. These diseases, in addition to the increase of degenerative diseases that occur primarily in those affluent societies where enzyme-deficient foods are chosen instead of raw fruits and vegetables, have sharply focused the public's attention on the inability of the medical community to improve nutritional health.

Many new approaches to the study of nutrition are emerging. The study of the cell has stimulated interest in the effect of genetics on nutritional need. The interaction between nutrition and genetics in human development is providing cures for some metabolic defects, such as phenylketonuria (PKU). This is a disorder in which there is a lack of the metabolic enzyme needed to handle extra amounts of the amino acid phenylalanine. This disorder results in mental retardation.

The effect of nutrition on brain development and behavior and on resistance to infection, stress, drug use, and environmental factors such as pollution are only a few of the new concepts being studied. Another area of interest is iatrogenic (doctor-caused) malnutrition, or nutritional disease resulting from treating a patient with drugs, surgery, or therapeutic diets. This is an extremely important area of research since the public is only now gradually becoming aware of the amount of harm caused by these treatments.

KEY POINT

In 1920 the first vitamin, vitamin C, was discovered. By 1940, several other water-soluble vitamins including B_1, B_2, B_3, B_6, pantothenic acid, and biotin had been identified.

Recognition by the Surgeon General in 1986 that nutritional factors play a part in the development or treatment of cardiovascular disease, hypertension, diabetes, and cancer, which are all leading causes of death among the general population, has led to extensive study of the role of these factors in preventing or curing degenerative diseases.

The Use of Food Supplements to Maintain Health

I previously spoke of the accepted definitions of nutrition. Simply stated, nutrition is the science of nourishing bodies. It ought to be obvious to everyone that a diseased person needs good nutrition. In fact, it can be argued that sick patients require very specific nutritional assistance to give their bodies the additional nutrients they need to combat the illness. However, that position evokes strong objections from established special-interest groups that long ago staked out disease as their private domain.

Over the next 10 to 15 years, you will see increased efforts to gain control over the practice of clinical nutrition and the industry of food supplementation. In fact, since the early 1980s many privately owned nutritional-supplement manufacturers have been bought out by large pharmaceutical corporations.

Until recently and despite such a clear-cut and logical definition of its area of concern or involvement in the health field, the practice of applied clinical nutrition has struggled to gain widespread respectability within the medical community. The scientific community has attempted to incorporate it within the broad parameters of medically related health professions, such as nursing, pharmacy, physical therapy, and hospital administration.

However, like physical therapy, nutrition has not flourished or developed under this tutelage. There are three reasons for this:

• Nutrition and physical therapy provide services that lessen the involvement of physicians, or detract from their image, or from their incomes.

- Many times these two areas of specialization are not generally familiar to physicians. For example, primary-care MDs have a maximum of 2.5 hours of formal nutritional training. Unless these doctors have been trained concerning their role in the healing process, it is not likely the patient will be referred for these services.

- Because these two areas of discipline have been under the wing of medicine, they have been unable to develop a philosophy of their own. This is particularly true of nutrition.

The general public, however, has long seen the innate wisdom of practicing good nutrition as a means of regaining health, in concert with the efforts of the medical practitioner. Unfortunately, the general public has had neither the physiological nor the biochemical educational background necessary to understand the action of specific nutrients and apply them logically for their own benefit. Neither have they had competently trained practitioners of nutrition to turn to for help. That statement will be vehemently contested by the American Dietetic Association and its members. While licensed dietitians are competently trained, they are not generally available to the general public, unless a patient is referred by a medical doctor. But much of their advice as members of the American Dietetic Association is suspect. The ADA received $3.2 million in 1995 from Coca-Cola, M&M Mars, McDonald's, Sara Lee, the Sugar Association, and the National Livestock and Meat Board, and its publications and practitioners often reflect bias in favor of these entities. Which brings us back to the issue concerning the development of nutrition under medical control.

To fill this public void, everyone and anyone remotely connected to the healing arts has been able to step in and begin dispensing nutritional guidance to anyone who wishes to pay for the service. This situation is like a double-edged sword: it cuts both ways. First, to direct the nutritional care of others in this country is to practice in a gray legal area. Due to the lack of proper education and licensing regulations, a great deal of fraud and quackery has been perpetrated on an unsuspecting and gullible American public, a fact that the medical establishment has made sure is constantly brought to our attention. On the other hand, this situation has kept nutrition in the public domain, not buried in the archives of medical science to be used at the medical establishment's convenience and contrivance, and has, so far, kept nutritional supplements and food from becoming prescription items.

Out of this confusion gradually came a few knowledgeable, well-intentioned spokespersons for the cause of nutrition in this country. Names like Royal Lee, Adelle Davis, Carlton Fredericks, and Linus Pauling began to give credence to the practice of nutrition as an alternative to medicine. More importantly, they gave us a glimpse of the profound idea that nutrition should be used to prevent illness and disease as opposed to medicine, which seeks to cure disease once it is found. This was an idea that the public, and, privately, even the medical community, could accept and eagerly embrace. Many doctors, of differing disciplines, began incorporating nutritional supplements into their practices. However, many new problems were encountered at this point. The most important, and most insidious, was the lack of philosophical discipline for the doctor to understand and use.

After all, most healing disciplines rely on the use of symptomatology for their implementation. The most obvious are homeopathy and medicine, which rely on the relief of symptoms in their quest for magic bullets. It was only natural that doctors would develop the same type of methodology in practicing the art of nutritional supplementation. Unfortunately, in the field of nutrition, this type of thinking leads only to confusion and lack of reproducible results. Take the symptom of hyperirritability, or just plain "nerves," for example. What nutrient or food supplement should be used?

- Vitamin B for lactic acid assimilation?
- Vitamin D for increased calcium utilization?
- Vitamin E to relieve anoxia?

- Vitamin F (fatty acids) to diffuse calcium throughout the body or to relieve hypothyroidism?
- Potassium to balance the resting membrane potential of the cells and relieve autonomic nervous system imbalances?
- Vegetables and whole grains to reduce refined sugar intake?
- Calcium to relieve a deficiency and feed the nervous system?
- Is the patient in a state of acidosis, requiring alkalizers for relief?
- Do we use distilled water to reduce the ionization of the body fluids?
- Is this a vitamin B_{12} deficiency?
- Is this an iodine deficiency?
- Is this a deficiency of sex hormones?
- Is it a matter of essential hypertension and does it require medical treatment?
- Or do we simply recommend tranquilizers or muscle relaxers?

Beyond symptomatology, do we have clinical tests that will give the answer? Not sophisticated, expensive laboratory procedures, but relatively inexpensive, reliable procedures the doctor can incorporate into the office procedure? With a few exceptions, the answer is no. It is true that blood chemistry will provide some answers for the nutritionist, like recognizing anemia and monitoring its correction. But blood chemistries have not been the boon to nutrition that they are to medicine, primarily because blood chemistries advise the doctor of developing pathology and not of physiological deviations before they become pathological. By definition, once blood chemistries are altered, this is clear evidence that the body has exhausted its ability to maintain health. Therefore, this is not an early-warning sign of a disease that can be prevented.

The evolution of nutrition to an exact science is filled with attempts to provide the practitioner with an instrument or means to know exactly what nutritional supplements this patient needs, as opposed to what is good for the general population in most cases. Most of these attempts have failed or have been removed from the market by the FDA. One need only recall the

Microdynameter, the Black Box, and the Endocardiograph. The Neuro-dermatron is now in vogue, as is dark-field microscopic examination of the blood. None of these methods have ever been fully convincing and therefore they have not been widely accepted or approved. This explains why health or nutritional questionnaires are still being used, particularly now that they can be computerized. But they bring us full circle, back to the beginning again—symptomatology. Why does vitamin C supplementation help some people and not others?

- How much do you use to prevent colds, how often, and when?
- Should it be from a natural source or synthetic, or a mixture of both?
- If you are using megadosages, are you practicing nutrition or medicine?

Why is it that medicine has been so successful in unraveling the most complex biochemical mysteries of the body and advanced their science so far in the last 50 years, yet nutrition, which by comparison is so fundamental in concept, has floundered in relative ignorance? Medicine and nutrition used to be one and the same. Physiology and anatomy were studied together (normal function) and medicines were food or herbs. The effort to restore normal function using natural methods was the rule. However, natural methods were not always successful at controlling disease, and in emergency situations and for treating advanced or chronic disease they were seldom effective.

So medicine set out on a new course and advanced its methods of scientific investigation. Now it is time for nutrition to advance itself by studying food and how the body uses it to maintain normal function, and leave the study of the effect of isolated chemicals on the body to medicine.

Nutritional Objectivity

Nutrition has traditionally been practiced through the recommendation of nutrients matched to the patient's symptoms. For example, if you have a cold, take vitamin C. If that doesn't work, add some vitamins A and D or adrenal

ENZYMES: *The Key to Health*

glandular support. If you are constipated, use a natural herbal laxative, and if that doesn't work, add pancreatin with ox bile for better digestion and maybe lactobacillus acidophilus for colon hygiene, and so on. This is a pharmaceutical or magic bullet approach that seeks relief of symptoms but ignores the source of the problem.

While these applications have been successful to a certain degree, they have not produced consistent results because such an approach fails to recognize the individuality of the patient. No two patients are alike. We each have been born with different genetic strengths and weaknesses, and the way we have treated our bodies has given us different states of vitality and recuperative capabilities.

Nutritional objectivity means that we should not be attempting to apply specific nutritional remedies to the population as a whole as magic bullets for symptoms. We should examine each individual and supply specific nutrients when needed to help the body meet increased nutritional demands during periods of stress.

Roger J. Williams, Ph.D., was professor of biochemistry at the University of Texas from 1940 to 1963 and director of its Clayton Foundation Biochemical Institute, where more vitamins and their variants have been discovered than in any other laboratory in the world. In his book *You Are Extraordinary*, he said, "There is no average person! We as individuals cannot be averaged with other people. Inborn individuality is a highly significant factor in all our lives—as inescapable as the fact that we are human."

What good does it do a person to know that on average, adult males over the age of 50 require 5,000 IU of vitamin A every day? Or that females in the

KEY POINT

Often patients do not seek nutritional help until a chronic degenerative disease has become evident or they are not satisfied with the relief pharmaceutical drugs are providing. But it is a little late for rose hips when you have just fallen off a 10-story building.

same age group require 8,000 IU? These are the recommended daily allowances (RDA). Yet the National Health and Nutritional Surveys (NHANES) in the 1970s found that *less than 1% of the older population had low vitamin A levels despite the fact that 50% of those tested had vitamin A intakes of less than two-thirds of the RDA!* The information is useful to science but has little or no validity when applied to an individual for a specific problem. A specific individual may require twice that amount or half that amount depending on the dietary demands being made by his body at any particular time in life.

Science employs scientific method, but it is very difficult to find a practical definition of this method, or where it originated. It appears to involve controlled conditions of experiments and reproducible results every single time the experiment is performed. But almost every recognized parameter of scientific method ignores the fact that human beings cannot be standardized.

In other words, we may put a "standardized" chemical into 1,000 people and measure the physiological effects, but each of those 1,000 people has a different set of parameters influencing the maintenance of health within the body. The results are therefore not scientific, but only an average. Good for some, bad for some, and ineffective for others.

The double-blind crossover placebo trial, therefore, is a sterile model of scientific methodology applicable only to pharmaceutical testing. In other words, for a remedy to be scientific, every asthmatic sufferer must have his symptoms relieved by exactly the same measured amount of a substance. This implies that the stimulus that has caused the bronchioles to constrict and become mucus-filled is apparently unimportant to the scientific understanding of the condition. Remember that the diagnosis of asthma is made by symptoms only, not by measurable objective testing.

You and an acquaintance of yours who also has asthma may both be exhibiting the same symptoms, but the problem may be different in each body. Therefore, since the human subjects involved in the healing process cannot be standardized, a better system is needed for evaluating herbs and whole-food substances. Instead of finding a remedy for a clinical symptom entity that will be beneficial for *everybody* (one word), we should concentrate our efforts on finding the solution for *every body* (two words). Clinical outcome studies are a more sensible answer.

Double-blind pharmacology studies to determine nutritional supplement usage are not of practical value in the practice of clinical nutrition. Yet we find over and over that nutrition attempts to use medical studies and medical methodology to apply nutrition, and it doesn't work consistently.

It must be remembered that there is an intelligence in living things. Because there is this intelligence, cellular activity is predictable. It can be relied on to give predictable information when examined based on its particular needs at the time of study, not based on the properties of the drug or chemical being applied to it.

Healing is an art as opposed to a science. The dictionary compares science to art: "A *science* teaches us to know, and an *art* to do, and the more perfect sciences lead to the creation of corresponding useful arts. Art is knowledge made efficient by skill."

We are often told that information is invalid if it was gathered before the discovery of the DNA/RNA molecule in the 1970s. The argument is that knowledge gained from genetic studies has completely revolutionized science. But human biochemistry is still working on the same principles as it was 2,500 years ago. True, some minor changes have occurred in our bodies and in our diets, but none in the laws that govern human biochemistry and physiology. We have only a more in-depth knowledge of the processes involved. Fifty years from now, our present knowledge will appear primitive.

KEY POINT

Healing is an art as opposed to a science. The dictionary compares science to art: "A *science* teaches us to know, and an *art* to do, and the more perfect sciences lead to the creation of corresponding useful arts. Art is knowledge made efficient by skill."

Unless we use a comprehensive method of determining the struggle the body is having to maintain health, we cannot expect to address and correct the root cause of the symptoms. Such an evaluation would not only consider nutritional status but must also determine the ability of that body to digest and assimilate nutrients.

Otherwise, supplying all the correct food or nutritional supplements will not solve the problem, and drugs or isolated chemicals (not food) must be used. Technically, that is the practice of medicine and not nutrition.

There are no magic bullets for every symptom in nutritional supplementation as there are in medicine. Individual nutrients cannot be matched to specific physiologic dysfunctions for the entire population. Again, the individual must be taken into consideration.

Everyone knows (or should know) that every drug will help some bodies, will not be effective in other bodies, and will harm still other bodies. Theoretically, it will not harm too many bodies, but clearly it is a numbers game. Such a philosophy totally ignores what caused the condition. The cause of the asthma, for example, may be different in *every body*.

Before you can understand enzyme nutrition, we have to decide what is health and what is disease. Only licensed health care practitioners can treat disease and administer sick care. Fortunately, you do not need a license to practice nutrition and direct your own health care. You don't even need an education, or even be able to read and write—everybody eats. But we need to define our terms so we (you and I) can understand each other.

It is not necessary to have any particular scientific background to understand the uses of enzymes. However, you need a working knowledge of the terminology or definitions of some of the basic terms used throughout this book. I've included some important definitions in Table 1 so you and I will have a common understanding of our meaning and intent in this work.

TABLE 1

ESSENTIAL DEFINITIONS

NUTRITION	vs.	PHARMACOLOGY
The science of food and the processes by which the organism ingests, digests, absorbs, transports, utilizes, and excretes food substances. *Webster's Third New International Dictionary of the English Language,* G. C. Merriam Co., Chicago		The science that deals with the study of drugs in all their aspects. *Webster's Third*

FOOD	vs.	DRUG
A material consisting of proteins, carbohydrates, fats, and other substances (minerals, vitamins, and enzymes) that are essential for an organism to sustain growth and repair and to furnish energy for all activity of the organism. *Webster's Third*		A substance, other than food, intended to affect the structure or function of the body of a human or other animal. *Webster's Third*

HEALTH	vs.	DISEASE
A normal condition of the body and mind, with all parts working normally. *Dorland's Illustrated Medical Dictionary,* 23rd edition, W. B. Saunders, Philadelphia		Disease is disturbed function ... out of time, out of tune, as well as disordered structure. Wm. S. Boyd, *Boyd's Pathology,* 6th edition, Philadelphia, Lea & Febiger

INTERNAL ENVIRONMENT

The environment in which each cell lives is called the internal environment, or the extracellular fluid, which surrounds the cell. It is from this fluid that the cells receive oxygen and nutrients and into which they excrete wastes.
Claude Bernard, French physician, 1859

The necessity of maintaining the internal environment relatively constant is the single most important idea to be kept in mind while attempting to unravel and understand the functions of the body's organ systems and their interrelationships.
Vander, Sherman, and Luciano, *Human Physiology,* 4th edition, McGraw-Hill, New York

HOMEOSTASIS

The term homeostasis is used by physiologists to mean maintenance of static or constant conditions in the internal environment.
A. C. Guyton, *Textbook of Medical Physiology,* 7th edition, W. B. Saunders, Philadelphia

A tendency to uniformity or stability in the normal body states (internal environment or fluid matrix) of the organism.
Dorland's

HEALTH

The harmonious integration of cellular activities in the various systems to meet an ever-changing environment.
W. D. Harper, *Anything Can Cause Anything,* privately published

In the late 1960s, noted Canadian researcher Hans Selye, M.D., wrote a book in which he presented his findings on the effect of stress on the human body. Thirty-five years after its publication, *The Stress of Life* can still be found on the shelves of your favorite bookstore. His work earned Dr. Selye the Nobel Prize for medicine and the acclaim of all those in the healing arts. Selye proved that stress is not a vague or undefinable term used to indicate that we are unloved, overworked, and underpaid. Rather, Selye found that the body responds to any kind of stress, be it mechanical, chemical, or emotional, in a very specific and predictable way.

Selye called this response by the body the general adaptation syndrome, and this is schematically represented in Table 2.

The chart begins with the body in a state of health or normalcy. If a stimulus (stress) is applied to it that requires a change on the part of the body to maintain normalcy, then an alarm signal is sent to the brain, which produces a resistance reaction. If the stimulus is removed and was not so strong as to cause tissue damage, then we probably don't even notice. Many such reactions occur in our bodies every day as we automatically adjust our rates of respiration and heartbeat as well as countless hormonal and autonomic nerve responses to meet changes in our environments, both external and internal.

If the stimulus is continued, then the body's response must continue to resist its effect and a state of compensation is reached—that is, the parts of the body affected by the stimulus/stress must elicit aid from other tissues or, in most cases, begin using increased amounts of nutrients to maintain its heightened state of function. This situation will continue as long as the stimulus is applied and as long as the flow of nutrition is maintained and the waste products formed by the affected organ/tissue are not allowed to accumulate.

However, once the tissue becomes fatigued because there is a lack of support from related tissues or organs, nutrition, or waste removal or because the stimulus is simply too strong, we enter a state of exhaustion. At this point we begin

TABLE 2

THE GENERAL ADAPTATION SYNDROME

Nutrition	Stimulus Alarm and Resistance Compensation
Exhaustion and the Appearance of Symptoms	
Medicine	Disease Degeneration Death

to experience symptoms. Consider this the demilitarized zone, or no-man's land. Looking back at Table 2, you can see that it is placed between nutrition and medicine. The affected tissues are not exhausted to the point of tissue damage. Objective findings such as physical examination, blood tests, and X-rays at this point are still negative, so we have not entered the zone of disease. Yet health has not been maintained, and we have left the zone of nutrition.

The big question now is, which way do we move? Up to health or down to disease, degeneration, and eventually death?

Selye's findings present us with the realization that since the human body has specific functions designed to maintain normal function and therefore health, attempts at healing should be directed toward relieving the stress and providing nutrients for the body to use in its defense. Such efforts would unquestionably fall within the province of nutrition.

Unfortunately, Selye's work is not appreciated and applied as often and judiciously as it might be. Instead, it is much more commonplace to find drugs (both over-the-counter and prescription) being used to cover up the symptoms or early-warning signs of stress. Such stress long camouflaged and ignored leads most often to disease, not health and well-being. Of course, it should be obvious that a diseased patient requires good nutrition as well as dietary supplementation to meet the increased demands being placed on the body.

Dr. Selye ended his book with a statement that I present here for your consideration:

> Apparently, disease is not just suffering, but a fight to maintain the homeostatic balance of our tissues … Could it be translated into the precise terms of modern science? Could it point a way to explore whether or not there is some nonspecific defense system built into our body, a mechanism to fight any kind of disease? Could it lead us to a unified theory of disease? (Hans Selye, *The Stress of Life*, 1968, McGraw-Hill, New York)

KEY POINT

The role of nutritional supplementation is to meet the increased needs of stressed organs. It would be like calling in some temporary help to get through the busy season. The situation is temporary yet stressful to all those involved.

One Disease: Stress

The concept of one universal disease is not new. But the presentation of such overwhelming research validation is. I would like to ask you to define health. How should doctors decide when a person is diseased and when that person is healthy? It is a really simple question, and everyone thinks they know the answer. What is your answer? Is health the lack of any apparent symptoms?

Health is defined as a normal condition of the body and mind with all the parts working normally. That's pretty simple. Disease is something else: it's disturbed function, something that isn't working normally. Did you know there are no new body processes at work in disease that were not there in health? In disease, there are only normal functions that are going too fast or too slow, or are otherwise inappropriate—out of time with need.

Every symptom crisis is produced by either mechanical, chemical, or emotional stress that either is too strong or continues too long for the body to be able to adequately compensate. *Any* stimulus that threatens homeostasis has disease-producing potential. Therefore, any treatment designed to suppress unpleasant symptoms diminishes the body's ability to protect itself.

Medical research attempts to find magic bullets for each symptom complex, as if the disease were the primary problem. But this is not a book about practicing medicine and treating disease. This is a book about nutrition and nourishing the body to prevent disease. So we view the human body as a self-healing organism and maintain that the proper way to prevent disease is to study health and every factor that can influence it, favorable or not. In other words, disease is perverted health; it cannot be its own cause, nor can it be its own cure, and certainly not its own prevention.

Health is the harmonious integration of cellular activities in the various systems to meet an ever-changing environment, and disease is disturbed function … out of time, out of tune, as well as disordered structure.

We can therefore recognize disease as the inability of the body to maintain a constant "normal" environment against the various stimuli or challenges that are presented to it. The question then becomes how to recognize this inability before it becomes disastrous.

Symptoms

Symptoms are our first clue that something is wrong within the body. Having symptoms does not necessarily mean, however, that a disease process has begun. Any time that a person has a symptom, look at the internal environment and you will see that some cell, tissue, or organ is under a considerable amount of stress. It may be mechanical, chemical, or emotional; the source doesn't matter. The cell is unable to perform its function(s) properly because the stress is either too strong or has been there too long, and the cell either doesn't have adequate nutrition or has accumulated too much waste. In other words, its environment is not adequate. Perhaps it is not doing its share to keep the environment normal, and other cells are suffering.

The rest of this society of cells will help out all they can to maintain the internal environment, even to the point of some of them dying. The body will rob Peter to pay Paul to keep the whole alive and well. It might have to destroy some cells to get some nutrient like protein or sugar, but it will do it as required. Only when it can no longer maintain the internal environment do you become diseased. This can be measured by some objective means, like a blood test, for example.

Enzyme Nutrition Paradigm

We recognize that disease is an insidious process that begins with small alterations in body chemistry or function, which usually go undetected by the patient.

1. If these alterations are not corrected exogenously (by changing the diet), then the body must compensate biochemically (endogenously) to correct itself.

2. When the body's ability to compensate is exhausted, it must alter normal physiological functions in order to sustain itself. This process produces symptoms but usually not physical findings.

3. If the process continues uncorrected, then the altered body functions will produce changes in structure or tissue. These changes are referred to as pathology and produce objective or physical signs.

Speaking very generally, in the healing arts today we have two widely divergent methods of restoring health:

1. The medical or drug therapy approach, which dictates the response of the body by taking control of its biochemistry.
2. The nutritional or holistic approach, which attempts to supply the necessary nutrients needed by the body to heal itself.

The difference between the two is that medicine deals primarily, and best, with pathology and objective signs. By the use of measurable changes, it detects and arrests tissue changes with drug therapy. This accounts for its continued growth and success. It deals with reproducible or measurable situations that have been developed into a scientific discipline.

By its ability to measure objective signs and apply reproducible drug therapy and benefits, medicine has grown and prospered in this age of technology. Such advances as the electron microscope and CAT scan have moved medicine progressively forward toward the 21st century.

Nutrition, on the other hand, has not advanced itself during this same period. Because it attempts to prevent disease before pathology develops, it has not had a means of measuring objective signs. It has contented itself with subjective evidence for its application, which unfortunately is neither reliable nor reproducible.

Nutrition has, however, benefited from the explosion of knowledge in the medical field. Such pioneers as Hans Selye, the Shute brothers, and Linus Pauling have given nutrition an air of respectability. In fact, nutrition owes its continued existence as an alternative healing art to these explorers. These as well as many other pioneers were well disciplined in scientific technology and were attempting to apply concentrated chemicals to the finding of objective evidence with the hope of establishing cause and effect for the chronic degenerative diseases plaguing mankind.

There is an enormous difference between nutrition and medicine for which there is a political battle being waged for control of natural healing substances. And as you can see, if you wish to practice clinical nutrition today, you have no alternative but to accept the work of medical researchers and apply their findings by supplementing crystalline-pure substances, such as ascorbic acid.

Table 3 will help us understand the fundamental differences between the practices of medicine (pharmacology) and nutrition.

Fortunately, there is a missing link in this progression from health to exhaustion and disease. It was discovered by Dr. Edward Howell in the 1930s. Dr. Howell realized the following:

1. Food is composed not only of protein, carbohydrates, fats, fibers, vitamins, and minerals, but also enzymes.

2. All living things contain enzymes; in fact, life cannot exist without them. Enzymatic activity is responsible for every biochemical reaction that occurs in living matter.

3. Enzymes are not just inorganic catalysts; they are biochemically active and therefore can be exhausted and become deficient in the human body. They are essential nutrients.

Dr. Howell recognized that food enzymes not only complete the picture for nutrition to become a science but also hold the keys to understanding chronic degenerative disease.

The food enzyme concept that Dr. Edward Howell began formulating in the first half of this century is detailed in his book *Enzyme Nutrition*. In the introduction to that book, Stephan Blauer states that Howell's conclusion is that "many, if not all, degenerative diseases that humans suffer and die from are caused by the excessive use of enzyme-deficient cooked and processed foods."

This opinion was certainly shared by two other farsighted nutritional pioneers whose investigations have been sadly neglected in this modern era of scientific enlightenment. I refer to Drs. Price and Pottenger.

Dr. Weston A. Price was a Cleveland dentist who traveled around the world in the 1930s studying the diets of people in isolated communities. Many, but not all, of these communities could be termed primitive cultures. He collected data

TABLE 3

FUNDAMENTAL DIFFERENCES BETWEEN
PHARMACOLOGY AND NUTRITION

Pharmacology	vs.	Nutrition
uses		uses
Drugs	**vs.**	**Food**
to		to
Treat disease	**vs.**	**Nourish the body and remove dietary stresses that challenge normal physiologic functions**
measured by		measured by
Symptoms **Case history** **Physical examination** **Laboratory tests**	**vs.**	**Symptoms** **Case history** **Physical examination** **Laboratory tests**
to identify		to identify
Pathology	**vs.**	**Challenges to normal physiology**
and establish a "working"		and establish a "working"
Diagnosis of disease	**vs.**	**Diagnosis of dietary stress(es)**
and		and
Treatment plan	**vs.**	**Treatment plan**

on the effect of diet on health, and in particular on dental health, and reported his findings in his book entitled *Nutrition and Physical Degeneration*.

His findings led him to the belief that there is a direct correlation between tooth decay and nutritional deficiencies. Generally, Dr. Price found that the natural indigenous diets of those groups studied were higher in both water-soluble and fat-soluble vitamins, as well as minerals such as calcium, than the average diet from more "civilized groups." Obviously, there was a noticeable lack of refined foods such as white flour, pasteurized milk, and convenience foods filled with preservatives and additives. This information correlates well with the results reported by Dr. Howell in his books and with that of Dr. Pottenger, who we will discuss next.

Price also reported that special attention was given to pregnant and lactating women. It can be pointed out that breast-feeding acts as a natural method of birth control and does not allow the mother to become pregnant soon after giving birth, thus allowing her to regain health. Breast-feeding was just beginning to be discouraged in more advanced cultures at this time, as baby formulas were becoming popular.

I have two books written by Francis Pottenger, M.D., in my library and have always been grateful for the knowledge contained in both. These are listed in the bibliography section under the Price-Pottenger Foundation. Pottenger's cat studies, like Price's observations, are classic in the field of nutrition, yet seemingly little appreciated today. Dr. Pottenger studied the effect of food on cats and followed their offspring for four generations.

Beginning in 1932 Dr. Pottenger studied more than 900 cats for over ten years, and found that only diets containing raw milk and raw meat produced optimal health. Cats on the all-raw diet had good bone structure and density, and enjoyed greater freedom from disease. They reproduced with ease and were gentle and easy to handle.

Cooking the meat or substituting heat processed milk for raw milk resulted in poor reproductive capabilities and physical degeneration, which increased dramatically with each generation. Birth defects and infant mortality of third generation cats were greatly increased. Skin diseases and allergies increased from an incidence of 5% in normal cats to over 90% in the third generation.

ENZYMES: *The Key to Health*

Bones became soft and pliable, calcium and phosphorus content diminished. Many cats suffered from hypothyroidism and most of the degenerative diseases encountered in human medicine. They died out completely by the fourth generation.

In short, what Pottenger observed in cats on deficient diets paralleled the human degeneration that Dr. Price found in tribes and villages that had abandoned traditional foods. Interestingly, this information also parallels that reported by the Surgeon General in the late 1980s.

In 1988 the Surgeon General of the United States released that office's second report to the American people. The first was in 1964 on the effects of smoking. This one, on health and nutrition, reported as its main conclusion that overconsumption of certain dietary components is now a major health concern for most Americans. The following quotes were taken from various sections of the Surgeon General's report:

> Foods contain nutrients essential for normal metabolic function, and when problems arise, they result from imbalance in nutrient intake or from harmful interaction with other factors.... This report reviews the scientific evidence that relates dietary excesses and imbalances to chronic diseases.... Diseases such as heart disease, stroke, cancer, and diabetes remain leading causes of death and disability in the United States. Substantial scientific research over the past few decades indicates that diet can play an important role in prevention of such conditions.... Good health does not always come easily. It is the product of complex interactions among environmental behavioral, social, or genetic factors. Some of these are, for practical purposes, beyond personal control. But there are many ways in which each of us can influence our chances for good health through the daily

choices we make.… For the two out of three adult Americans who do not smoke and do not drink excessively, one personal choice seems to influence long-term health prospects more than any other: What we eat.… Yet what we eat may affect our risk for several of the leading causes of death for Americans, notably, coronary heart disease, stroke, atherosclerosis, diabetes, and some types of cancer. These disorders now account for more than two-thirds of all deaths in the United States.… It is not yet possible to determine the proportion of chronic diseases that can be reduced by dietary changes. Nonetheless, it is now clear that diet contributes in substantial ways to the development of these diseases and that modification of diet can contribute to their prevention.

The message is clear: chronic degenerative diseases must be prevented. The report goes on to say that scientific studies over the past 30 years offer evidence that the incidence of chronic degenerative diseases can be lowered by dietary changes, and that the problem lies not in deficiencies but in dietary excesses.

An excellent example of the problems outlined in the Surgeon General's report was reported by Electa Draper in the *Denver Post* of May 17, 1998, under the title "Diabetes a Navajo Epidemic." In the Montezuma Creek and Monument Valley project area of southeastern Utah, home to 7,000 to 8,000 of the nation's estimated 240,000 Navajos, the incidence of diabetes in the patient population is estimated at 40% and increasing.

Twenty percent to 50% of the adults in many Native American Indian tribes have been diagnosed with diabetes, according to Indian Health Service reports. The actual incidence could be as high as twice that because, according to the American Diabetes Association, as a rule, for every diagnosed person, there's someone who has the disease but doesn't know it yet. The most prevalent type of diabetes, Type II or adult-onset diabetes, which is related to diet, was almost unheard of on the reservation before 1940. Now it is reportedly an epidemic and has been found in Navajo children as young as eight years old.

The blame for this alarming increase is traced directly to convenience stores, supermarkets, and fast-food chains from outside the reservation that have changed the landscape and the people in just a few generations. On a wall at the Montezuma Creek clinic, a 44-ounce soda pop cup, offered at one chain as "The Giant," is taped onto a board next to 55 sugar cubes. That's how much sugar the soft drink holds. In desert country, where many Navajos have to haul their drinking water that usually tastes terrible, people drink soda pop as if it were water.

When excess food and drink can always be obtained more readily and cheaply than fresh produce, and when modern occupations are not physically strenuous, the results of this genetic predisposition are rampant obesity and diabetes.

Consumption of alcohol, which is high in calories, aggravates and may even contribute to the onset of diabetes, but alcohol abuse is not the cause of the epidemic on the reservation. Prior to World War II, commodities such as white sugar, white flour, and lard had no part in Navajo life. It wasn't until the 1920s and 1930s that Navajos began taking in a standard American diet high in sugar, fat, and salt.

KEY POINT

Improved diet and good digestion can **prevent** chronic degenerative diseases.

Diabetes, if uncontrolled, can cause blindness, kidney failure, and vascular disease and may end in amputation of toes, feet, and even legs. Such extreme manifestations of the disease occur 8 to 10 times more often in the Navajo population than the general diabetic population.

Do not think that this is just a problem found on one Indian reservation. The Associated Press reported on February 12, 1997, that a high-starch, low-fiber diet raises the risk of diabetes. The report quoted a study reported in an issue of the *Journal of the American Medical Association* that said middle-aged women who eat a lot of white bread, potatoes, and certain other starchy or sugary foods are more likely to get diabetes. The risk is especially high among those who skimp on cereal fiber, according to the six-year study of 65,173 nurses.

The study involved Type II diabetes, by far the most common form of disease in the United States. It afflicts more than 14 million Americans. The researchers were led by Dr. Jorge Salmeron of Harvard Public Health, who previously reported similar findings in men.

Chapter 2

A Society of Cells

Having looked at the evolution of nutrition and its application in our society, let us next turn our attention to the object to which nutrition is applied—the human body. We will start by examining the simplest structural unit into which a complex multicellular organism can be divided and still retain the functions characteristic of life: the cell. Just as the atom is the smallest particle of an element that can enter into a chemical reaction, the cell is the most fundamental unit to which we can reduce the human body and still study it.

One of the unifying generalizations of biology is that certain fundamental activities are common to almost all cells and represent the minimal requirements for maintaining cell integrity and life, such as:

- Consistently demonstrating a desire to live.
- Taking food from the surrounding environment and excreting its waste back into that environment.
- Making energy by breaking down these organic nutrients and making new molecules from them.
- Reproducing themselves.

Thus, a human liver cell and an amoeba, for example, are remarkably similar. In fact, they look alike under an ordinary light microscope. Because they do not have a circulatory system, food can be diffused only over a very small area; therefore, all cells are about the same size. However, because of information locked in the genetic code of human cells and amoebae, the two will have very different destinies.

The amoeba does not require a partner to reproduce. It simply divides on its own by the process of mitosis. In other words, somebody made the first one, and it has been doing its own thing ever since. While it does have a life cycle, it never becomes specialized.

Cell Specialization

The development of all vertebrates begins with fertilization of the ovum by the sperm to produce a single cell. This cell divides to form two cells, each of which divides in turn, resulting in four cells, and so on. This process is called mitosis.

If cell multiplication were the only event occurring, the end result would be a formless mass of identical cells. Instead, these cells soon begin to differentiate and become specialized in their functions. Microscopic studies have identified about two hundred distinct kinds of cells. However, all human cells can be classified into four broad categories based on the functions they perform:

1. Muscle cells produce movement.
2. Nerve cells initiate and conduct impulses.
3. Epithelial cells absorb and secrete organic molecules and ions.
4. Connective tissue cells form and secrete various types of extracellular connecting and supporting elements.

Differentiated cells with similar properties aggregate together to form tissues. These specialized cells arrange themselves in various proportions and patterns and combine with other tissue types to form organs. Within the organs, the four types of tissue are arranged in sheets, bundles, tubes, layers, and strips, with each subunit serving an important role in the function of the organ.

Finally, the organs can be classified into ten organ systems (see Table 4), with each system performing the same tasks that an individual cell performs for itself to maintain its life. This organization of systems provides the basis for a doctor's case history when a patient seeks treatment for symptoms.

Having examined the development of the amoeba and human cells, let's now look at a common need they both have in order to perform the essential tasks necessary to maintain their lives.

The amoeba lives an independent life in its environment. Human cells, however, are capable of living only if they are in an appropriate environment. In

TABLE 4

TEN ORGAN SYSTEMS

1. Circulatory	6. Musculoskeletal
2. Digestive	7. Nervous
3. Endocrine	8. Reproductive
4. Immune	9. Respiratory
5. Integumentary (skin)	10. Urinary

other words, a specialized human cell is dependent on the ten organ systems to bring its environment to it. These ten organ systems exist only to provide an internal environment and act as a life-support system for each of the body's individual cells. Somebody's got to grow the food. Somebody's got to build the houses. Somebody's got to supply the electricity. Somebody's got to stand guard. Somebody's got to deliver the groceries. Somebody's got to pick up the garbage. Somebody's got to create energy, and so forth and so on. That's all these systems do, maintain the internal environment.

A single-cell organism (like an amoeba) can exist as its own entity. It can go find food on its own. When it finds something tempting, the single-cell organism surrounds the food and takes it into itself. It then performs some chemical magic and converts the food into energy, or uses it to repair itself. That single-cell organism can then spit the waste back into the environment.

But one of the individual cells in your body can't do that because it has become specialized. If we removed a liver cell, for example, from its internal environment and put it on the table, it would die. Your specialized cells all have to work together like a society, each doing its share to provide for the whole. Their environment must be maintained and brought to them so they can prosper. Just like in your own home, someone must carry in the groceries and take out the garbage, and someone must prepare the meals. The amoeba doesn't have to do that. It can forage for food and throw its garbage anywhere it falls—just like the guy driving down the road in front of you throwing his cigarette butts out his car window. However, each cell is responsible for its share in maintaining the environment, just like paying taxes. After all, someone has to clean up the road to maintain the environment for the rest of us. You are aware of your external environment, but your cells depend on an internal environment. Remember, the cell is counting on this environment. The cell gave up certain inalienable rights to be part of this society, and it's going to demand services. It expects the garbage to be picked up.

The Internal Environment and Homeostasis

From the time the cell was identified as the basic structural and functional unit of life, physiologists have recognized the importance of maintaining a constant internal environment. Claude Bernard, the famous French physiologist who introduced the term *internal environment (milieu intérieur)*, is chiefly responsible for this basic concept. The American physiologist Walter Cannon extended the concept of a constant internal environment and coined the term *homeostasis* (Greek *homoios*, like; Greek *stasis*, position) for the "steady state" conditions that are maintained by coordinated physiological processes.

ENZYMES: *The Key to Health*

The internal environment is a body fluid compartment and exists as a specific anatomical entity known as the *extracellular fluid (ECF)*. It is from this fluid that the cells receive oxygen and nutrients and into which they excrete wastes.

The ability of the body to digest, absorb, transport, and utilize food and eliminate waste is based on the relationship that exists between cells (the consumer) and the internal environment (grocer and trash hauler). Your health is very much dependent on the price being paid by the body to maintain a constancy of the volume, temperature, pH, and solute concentrations in the ECF.

The concept of the internal environment and the necessity of maintaining it relatively constant is the single most important idea to be kept in mind when trying to understand the functions of the body's organ systems and to explain the symptoms a person is experiencing.

I remember briefly studying the concept of homeostasis in college. It was a pretty simple straightforward explanation of how the body maintained equilibrium between its various systems and organs.

I was quite content to regard homeostasis as a state of constancy that the body maintained all the time. It was something I could depend on and take for granted. Certainly, as a chiropractor I would not be called upon to adjust patients who were in the process of losing control of their homeostatic functions. I know now that I missed one of the most important concepts in healing. Unfortunately, it was not until about 15 years later that I gave the subject any more serious thought.

While collecting clinical data on digestive complaints and correlating them to laboratory results, I came face to face with the ability of the body to maintain homeostasis. How was this accomplished since the blood must provide both the acidity for HCl production in the stomach and alkalinity to the duodenum for the pancreatic enzymes to work? It sent me scurrying to my medical library for an exact definition, and even back to Claude Bernard's original concept of the "Internal Milieu" in 1859.

FIGURE 1

SCHEMATIC REPRESENTATION OF TOTAL BODY WATER

(60% of Total Body Weight)

The following chart attempts to demonstrate that the blood is only a small part of the total body water and is used as a fluid transport medium carrying nutrients to the cells and picking up their waste.

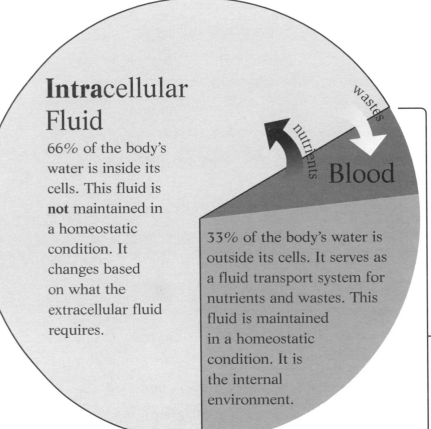

Intracellular Fluid

66% of the body's water is inside its cells. This fluid is **not** maintained in a homeostatic condition. It changes based on what the extracellular fluid requires.

33% of the body's water is outside its cells. It serves as a fluid transport system for nutrients and wastes. This fluid is maintained in a homeostatic condition. It is the internal environment.

wastes

nutrients

Blood

20% of the ECF is in the blood. The body strives to maintain the following constants in the blood:
- pH (acid/alkaline)
- volume (blood pressure)
- temperature
- concentration of dissolved substances such as:
  cholesterol
  glucose
  iron
  triglycerides
  hormones

Extracellular Fluid (ECF)

The environment in which each cell lives is called the internal environment. This environment is the extracellular fluid, which surrounds the cell. Twenty percent of this fluid is found in the bloodstream, and 80% in connective tissue.

It is from this fluid that the cells receive oxygen and nutrients and into which they excrete waste. Yet it must constantly be maintained within the very narrow limits of body temperature, pH, volume or water content, and concentration of dissolved substances such as sugar, cholesterol, and many more.

Guyton's Medical Physiology (7th edition) states: "The term homeostasis is used by physiologists to mean maintenance of static or constant conditions in the internal environment." It is a tendency to uniformity or stability in the internal environment or fluid matrix of the organism. This is the material that I skimmed over in school, taking for granted that the body would do this regardless of what I did.

I did not completely appreciate that the body will do whatever it must to maintain homeostasis! In other words, I should be able to somehow measure the early warning signs of exhaustion in the struggle to maintain normalcy.

By definition, blood chemistry studies are not the answer since the body has already lost the struggle by the time the blood test is beyond normal limits. As an example, recall that blood calcium and phosphorus levels remain normal in osteoporosis. This is accomplished by removing the minerals from storage in order to maintain homeostasis—at all cost. There are many other such examples of the body's willingness to stress specific organs in order to meet the requirements of the extracellular fluid. Obviously, the stressed tissues have increased nutritional requirements during these periods.

As depicted in Figure 1, the blood exchanges nutrients and wastes with the interstitial fluid (between and around the cells) which in turn exchanges nutrients and wastes with the cells.

Sixty percent of your body weight is water and is contained in three separate compartments:

1. The fluid inside the cells (intracellular fluid) accounts for two-thirds of the total body water.

2. The fluid outside the cells, the extracellular fluid (ECF), accounts for the other one-third of the total body fluid. It is this fluid, the internal environment, that must remain relatively constant in terms of acid-alkaline balance (pH), temperature, volume (the amount of water), and levels of dissolved substances (such as sugar, protein, cholesterol, iron, and so on) needed to nourish the cells.

3. The extracellular fluid is further divided into two compartments:

   • The blood is only 20% of the extracellular fluid.

   • The other 80% the ECF is found in the connective tissue and is called the *interstitial fluid*. This lies under the skin and in fact can be palpated for its consistency.

   • The contents of the blood and this interstitial or intercellular fluid are identical, except for their protein content. Additional protein is needed in the blood to hold water; otherwise it would leak out into the tissues and you would be aware of swelling, first in the ankles.

It is important for you to realize that the interstitial fluid is contained within the connective tissue of the body. In other words, there are no empty spaces within the body. The lumen of the digestive, respiratory, and urinary tracts are considered to be outside the body. And, just as the surface of your body is covered with a protective covering of skin, so are the walls of the digestive tract (beginning with your lips) and the respiratory tract and the urinary tract.

In the digestive tract this lining of skin, called *epithelial tissue*, is also protected from the digestive juices by a thick coating of mucous cells. These cells not only provide protection for the walls against the strong digestive juices, but they also provide lubrication for the passage of food and waste materials. We often forget that these cells must also receive nutrients and have their waste removed by the internal environment in order for them to function properly. When this does not occur, we suffer from such symptoms as heartburn, gastritis, and even ulcers. This can occur anywhere along the digestive tract—in the stomach; in the small intestine, especially the duodenum; and in the large intestine as in irritable bowel syndromes such as Crohn's disease and ulcerative colitis.

Mechanisms of Control

Let us now take a look at how the body coordinates all of the many cellular and systemic activities necessary to maintain health.

Implicit in life is control. Regardless of its level of organizational complexity, no living system can exist without precise mechanisms for controlling its various activities. Every one of the fundamental processes performed by any single cell must be carefully regulated. Information about all important aspects of the external and internal environments must be constantly monitored by receptors

FIGURE 2

THE MAJOR PATHWAYS OF CONTROLLING HOMEOSTASIS

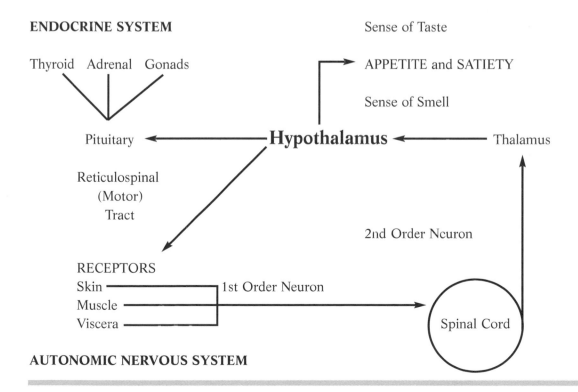

and sent to the brain. The brain then directs the nervous and hormonal systems to send instructions back to the various tissues and organ cells, directing them to increase or decrease their activities.

The body has two major control systems for maintaining homeostasis: the autonomic nervous system and the endocrine (or hormonal) system. Both of these systems receive signals and direction from the hypothalamus gland. Utilizing input from the various parts of the body and the state of the extracellular fluids, the hypothalamus continuously directs these systems in an effort to maintain homeostasis.

The hypothalamus contains centers for monitoring and regulating the temperature, volume, and concentration of solutes of the extracellular fluids. Further, it is linked by nerves to most parts of the brain.

When you look at histological slides of a 16-day-old fetus, you see that it has three layers identified as an outside, an inside, and a middle. After about three weeks, the outside layer (ectoderm) becomes the skin, the nervous system, and the brain. The inside layer (endoderm) becomes, for the most part, the organs. The middle layer (mesoderm) essentially becomes muscle and connective tissue, the glue that holds you together. As the fetus continues to develop, these cells begin to specialize, and the specialized cells begin to congregate together to form bundles, sheets, and tubes. Then they begin to develop into organs.

What this means is that we can really only complain about three basic areas of symptoms: skin, muscle, and organ. By looking at Figure 2 you can probably guess what they are. The brain communicates with the body via your spinal nerves. Each spinal nerve has receptors located in the ectoderm (skin and nerves), mesoderm (muscles, tendons, and ligaments) and the endoderm (organs and other tissues). When the complaint is organ dysfunction, we have only to ask what function that organ performs to maintain the internal environment. Is that function being performed too fast or too slow, or is it out of time with other body functions?

The Hypothalamus

It is interesting to note that the hypothalamus is the only part of the brain that is not protected by the blood-brain barrier. That barrier exists to protect the delicate tissues of the brain from changes in the extracellular fluids. But the hypothalamus is exposed to the extracellular fluid or internal environment.

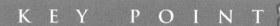
The body has receptors located in the skin, muscles, and organs of the body. These develop together in the embryo and grow from the spinal cord to the tissues. In other words, signals from any of these areas pass through neurons on their way to the spinal cord. The signals from these receptors then coalesce in the cord and take one of three pathways to the thalamus in the brain. The information is then passed to the hypothalamus gland and the postcentral gyrus, where the body makes decisions on how to respond to these signals. The answers are then conveyed back to the various tissues, which react accordingly. The response is carried over the reticulospinal tract back to the skin, muscles, and organs. Of particular importance here is that *all three tissues* are stimulated to respond.

An excellent example of these interrelated associations is seen in the very common occurrence of shoulder pain that is actually caused by gallbladder dysfunction. In the developing embryo, the diaphragm, which separates the chest from the abdomen, is actually found in the middle of the neck. As the embryo develops, the diaphragm is pushed downward, carrying its nerve connection from the neck (spinal nerves C3 to C5) with it. As the embryo develops and forms its spine and ribs, nerve connections from the thoracic spine are also made to the diaphragm. These nerve connections form communication between the diaphragm with both the thoracic and abdominal organs as well as the muscular and connective tissues of the shoulder, thorax, and abdomen.

Therefore, nociceptive (pain) impulses originating in either the pleura (above the diaphragm) or the peritoneum (below the diaphragm) can be responsible for muscle contraction in the cervical spine and shoulders.

The Importance of Understanding Negative Feedback

All the homeostatic control mechanisms of the body operate by a process of *negative feedback*. The feedback is, in effect, an informational signal that tells the monitoring tissue (the brain, for example) how well it is doing at establishing or maintaining some variable at the desired level. For instance, if the oxygen levels in the body fluids become too low, this information is fed back through nervous or hormonal stimulation to the mechanism for controlling oxygen, which automatically returns the oxygen to a higher level. The feedback is termed negative because it counterbalances, or negates, the change. A good example of this occurs in elderly patients having their blood pressure checked. Hospital studies confirm that it is quite common for blood pressure to rise because of apprehension (*American Journal of Hypertension*, volume 98, issue 11, pages 203–207). A prescription is then written to lower it. By the time the prescription is filled, the pressure has returned to normal. Now each time the medication is taken, the body counteracts it by raising the blood pressure back to normal, and a never-ending cycle is begun.

Before we proceed, it may be helpful to review some of the body processes controlled by the hypothalamus. Mainly through studies of laboratory animals, the hypothalamus has been found to be concerned with the regulation of peripheral autonomic nervous system discharges accompanying behavior and emotional expression. Among hypothalamic functions are the regulation of body water and electrolyte concentrations, temperature control, and regulation of feeding activities and the metabolism of fats and carbohydrates. It also manufactures the hormones of the posterior pituitary gland and controls secretion by both the posterior and the anterior pituitary, and thus has an important role in regulating endocrine functions and in maintaining healthy sexual behavior and reproduction.

Additionally, the hypothalamus controls both appetite and satiety. It is linked directly to your sense of smell and taste. It motivates you to eat what the extra-cellular fluid requires, and to reject what it does not need by making certain foods and drinks taste or smell good or bad. We generally crave what we cannot adequately digest and deliver to the hypothalamus via the bloodstream.

Appetite is the longing for a preferred food substance and is associated with conditioned reflexes such as salivation and secretion of stomach juices. While food preferences are determined by previous experience, genetic factors, and dietary deficiencies, specific hungers often occur when the body needs certain substances. Human infants and primitive peoples usually prefer foods that are best for them. In one study, where babies were allowed to select their own foods, an infant with rickets at the beginning of the study cured himself by selecting large amounts of cod-liver oil containing vitamin D.

Hunger pangs can occur as the result of strong stomach contractions, although the hunger drive can clearly operate in the absence of these contrac-tions, which may be only a by-product of the physiological state of the individ-ual. For example, lab animals without stomachs exhibit a desire for food more often than do normal animals. A lowered blood-sugar level apparently is an important factor in creating desire for food. This factor operates through cen-ters in the hypothalamus, which contains more blood vessels than any other part of the nervous system and receives circulatory, chemical, sensory, and neural stimuli.

The hypothalamus also plays an important role in the control of other drives, such as thirst. In the case of thirst, the intensity of the drive is related to the water deficit. Thus, a dehydrated individual will consume enough water to replace his water deficit within about 30 minutes. It is thought that concentra-tions of salt and other chemical substances in the body fluids, as well as the total amount of fluid present in the body, help to determine thirst.

Nausea is a disagreeable sensation in the pit of the stomach that may or may not be associated with vomiting and that is carried by both the vagus nerve and by sympathetic nerves. A vomiting center in the reticular formation of the medulla receives impulses from a chemoreceptor trigger zone located above the area postrema of the brainstem. This zone is the central site of action of the emetic drugs, which cause vomiting.

The sensation of *air hunger* is a result of excess carbon dioxide accumulation in the lungs. Although oxygen is a constant requirement of the body, the body cannot respond directly to oxygen lack. However, as oxygen is used up, carbon dioxide collects in the lungs, causing great discomfort unless the reflex breathing mechanisms are able to again substitute oxygen for carbon dioxide. When oxygen starvation occurs in an atmosphere where carbon dioxide has not increased (for example, at high altitudes), a kind of intoxication results. The individual may undergo memory impairment and paralysis or may shout or burst into tears. At the same time the person feels confident of her abilities and does not realize the seriousness of this condition. There is evidence that partial oxygen starvation brings out emotional reactions that are normally held under voluntary control.

Consciousness of *fatigue* impels human rest. Everybody is aware of how strong the need for sleep can become. As a result of prolonged exercise, the chemistry of the blood is altered in several ways. An elevated concentration of lactic acid in muscles presumably stimulates the nervous system directly or activates certain receptors. In sleepiness, it may be that nerve and brain centers are directly stimulated by chemical conditions in the body. The story of fatigue is complicated by the fact that it sometimes seems to result not from physical exertion but from boredom, worry, or frustration.

Obesity has been observed in individuals with tumors of the pituitary that are encroaching on the hypothalamus gland. Destruction of the satiety center in rats, cats, and monkeys causes overeating (hyperphagia) and obesity. Destruction of the feeding center results in a cessation of feeding (aphagia).

It has been suggested that when the desire for food is satisfied, the satiety center inhibits the feeding center; if the satiety center is destroyed, control over feeding is lost. The satiety center concentrates glucose and responds to increases in blood glucose levels with increased electrical activity. Drugs that reduce appetite, such as amphetamines, also increase the electrical activity of the satiety center. It is generally believed that the activity of the satiety center is also influenced by input from peripheral receptors. There is some reason to believe that feedback mechanisms developed early in life may permanently affect feeding habits. It has been found, for example, that overfed infants develop a greater number of fat cells in fat deposits than normally fed infants. The number remains constant—weight-reducing regimens reduce the amount of fat in the cells but not the number of cells. This may explain why overfed infants tend to be overweight all their lives.

The hypothalamus contains centers for regulating body temperature and the volume of body fluids, among others, and is linked by nerve circuits to almost all parts of the brain.

The "Master Endocrine Gland"

Utilizing input from other parts of the brain and information received from the blood passing through it, the hypothalamus continuously regulates almost the entire endocrine system.

KEY POINT

The cell is like a citizen within the physical community. It is the lowest common denominator. Each cell belongs to a group (an organ). Beyond that, each organ is regulated under strict guidelines by a governmental entity—in this case, the hypothalamus. Each citizen cell pays its taxes and demands services in return. It requires that the trash is cleaned up and the environment surrounding it is maintained.

It is well known that the hypothalamus synthesizes and secretes hormones (called *releasing and inhibitory factors*) that control the secretion of the hormones of the anterior pituitary gland. Specialized neurosecretory cells in various parts of the hypothalamus synthesize the releasing and inhibitory factors. These factors are absorbed by the bloodstream and carried to the anterior pituitary. They are transported to the pituitary by a direct vascular link between the two glands. The pituitary is connected to the hypothalamus by a stalk, and blood vessels arising from capillaries in the hypothalamus pass down the stalk into the anterior pituitary.

Four of the hormones of the anterior pituitary gland control the functions of other endocrine glands, namely, the thyroid, adrenal cortex, ovary, and testis. Prolactin governs milk production, and growth hormone causes general body growth, among other things. It should be apparent, then, that the hypothalamus, by regulating the secretion of the hormones of the anterior pituitary, exerts considerable influence over a large part of the endocrine system.

Ordinarily, these interactions contribute to normal development and health. Disturbed emotional states, however, may have the opposite effect. For example, poor growth in children deprived of normal parental affection (a condition called *deprivation dwarfism*) is apparently in large measure a consequence of a deficiency in the secretion of hypothalamic releasing factors, including the growth hormone-releasing factor.

The Endocrine Glands and Their Secretions

Hypothalamus: Secretes the following releasing factors that cause the anterior pituitary gland to increase its own hormone production, which stimulates the other endocrine glands.

- *Thyrotropin-releasing factor* (TRF)
- *Corticotropin-releasing factor* (CRF)

- *Follicle-stimulating-hormone-releasing factor* (FRF)

- *Luteinizing-hormone-releasing factor* (LRF)

- *Growth-hormone-releasing factor* (GRF)

- *Growth-hormone-inhibitory factor* (GIF)

- *Prolactin-inhibitory factor* (PIF)

- *Prolactin-releasing factor* (PRF)

- *Melanocyte-stimulating-hormone-releasing factor* (MRF)

- *Melanocyte-stimulating-hormone-inhibitory factor* (MIF)

Anterior pituitary gland:

- *Thyroid-stimulating hormone* (TSH)

- *Adrenocorticotropic hormone* (ACTH) stimulates the adrenal cortex.

- *Follicle-stimulating hormone* (FSH) stimulates follicle growth and maturation in females. Influences endometrial changes.

- *Luteinizing hormone* (LH) produces sex hormones.

- *Prolactin* stimulates breast milk secretion.

- *Growth hormone* (GH) increases growth of tissues.

- *Melanocyte-stimulating hormone* (MSH) produces skin pigment.

Thyroid:

- *Thyroxine* (T4) and *triiodothyronine* (T3) stimulate metabolic rate.

- *Calcitonin* influences blood calcium levels.

Parathyroid: This gland controls calcium and phosphorus levels.

Posterior pituitary gland: Specialized centers in the hypothalamus manufacture these two hormones that pass down axons of the pituitary stalk to nerve endings in the posterior pituitary gland, where they are stored. These hormones can then be released by nerve signals that arise in the hypothalamic gland itself.

- *Antidiuretic hormone* (ADH) influences water levels in the body.

- *Oxytocin* initiates uterine contraction during birth and allows for the flow of breast milk.

Adrenal cortex:

• *Cortisol* assists the body's response to stress.

• *Aldosterone* controls sodium and potassium in the body.

Kidneys:

• *Renin-angiotensin* regulates blood pressure.

• *Erythropoietin* is involved with red blood cell production.

• *1.25 dihydroxyvitamin D_3* is involved with blood calcium balance.

Adrenal medulla:

• *Epinephrine* and *norepinephrine* are released in response to stress and influence cardiovascular function. The hypothalamus controls the secretion of epinephrine and norepinephrine by the adrenal medulla via a direct nerve pathway. These fibers pass down the spinal cord and connect synapse with preganglionic neurons whose fibers lead to the adrenal medulla.

• *Androgens* support female sexual functions.

Islets of Langerhans of the pancreas:

• *Insulin* lowers blood sugar levels.

• *Glucagon* raises blood sugar levels.

Ovary:

• *Estrogen* supports female reproductive function.

• *Progesterone* supports sexual growth and development.

Testis:

• *Testosterone* supports male reproductive functions.

Gastrointestinal tract: These hormones regulate digestion and assimilation.

• *Gastrin*

• *Secretin*

• *Cholecystokinin*

• *Gastric inhibitory peptide*

• *Somatostatin*

Liver:

• *Somatomedin* stimulates bone growth.

Thymus:

• *Thymosin* orchestrates white blood cell development.

Pineal gland:

• *Melatonin*

Placenta: These hormones maintain pregnancy.

• *Chorionic gonadotropin*

• *Estrogens*

• *Progesterone*

Controlling the Autonomic Nervous System

We mentioned previously that the body has two main systems with which to control and integrate its various organs and tissues to maintain homeostasis. With the first being the endocrine system, the second of these is the autonomic nervous system. While not technically correct, one could call this system the automatic or subconscious nervous system. This system directs all of those necessary body processes that you and I would forget. After all, if there are times when you cannot remember where your keys are, it is only logical to assume that on occasion you would forget to breathe or keep your heart beating.

The nervous system allows the various parts of the body to communicate with each other, and it has two divisions: the

KEY POINT

The autonomic nervous system acts much like a city council—establishing rules and regulations for others to follow, directing the actions of others to benefit the community as a whole. The endocrine system works along with the autonomic system by acting as the city works department—following through with the direction of the city council and ruling on other important matters at hand. Presiding over these actions is the hypothalamus acting much like a mayor would—carefully balancing the two powers and guaranteeing the citizens will get what they ask for.

central nervous system (the brain and spinal cord) and the *peripheral nervous system* (the nerves that run from the central nervous system to the tissues and organs of the body).

The peripheral nervous system is further divided into the *afferent division*, which carries information from the tissues to the brain, and the *efferent* or *autonomic system*, which carries the brain's answer or instructions back to the tissue or organ.

Remembering that the body must maintain an internal environment within very narrow physiological limits, it can be appreciated that the autonomic system would also have two parts, the *sympathetic division* and the *parasympathetic division*.

It is beyond the scope of this book to give a complete description of the autonomic nervous system, and that is not necessary for our discussion. Just as we have gas pedals and brakes in our automobiles, the brain needs one set of controls to increase activity in an organ/tissue and separate controls to decrease activity in the organs/tissues.

It would be convenient if the body used either the sympathetic or parasympathetic as a gas pedal and the other as a brake, but that is not true. Both the sympathetic and parasympathetic systems can stimulate some systems and sedate others. Fortunately, there is no need for a more detailed description here; however, Table 5 summarizes the effects of each system on the various organs.

The concept of sympathetic and parasympathetic controls is an intriguing one when you consider that they are both dependent on receiving proper nutrition and having their waste removed in order to function normally, just like every other part of the body. In other words, imbalances between the two systems can occur.

There seems to be the misunderstanding that one system can become totally dominant over the other and cause total body symptom patterns. However, this happens only in extreme or emergency situations, such as in "fright, flight, or

ENZYMES: *The Key to Health*

TABLE 5

EFFECTS OF THE AUTONOMIC NERVOUS SYSTEM

Sympathetic Effect	Organ	Parasympathetic Effect
Dilates pupils	Eyes	
Dries	Lacrimal glands	
Dries	Nasal sinuses	
Increases secretion of enzymes in saliva	Salivary glands	Increases secretion of watery element of saliva
Increases secretion	Thyroid	Decreases secretion
Increases heart rate and force of contraction	Heart	Decreases heart rate and force of contraction
Relaxes muscles	Bronchi	Tightens muscles
Increases respiration rate	Lungs	
	Esophagus	Contracts
	Cardiac valve	Constricts and relaxes
Decreases digestive secretions	Stomach	Increases digestive secretions
Contracts	Pyloric valve	Relaxes
Relaxes muscles and constricts sphincter	Gallbladder	Relaxes sphincter and contracts muscles
Decreases secretion	Pancreas	Increases secretion
Contracts muscles	Spleen	
Increases glucose levels Increases protein levels Reduces secretion of bile	Liver	Reduces protein levels Increases secretion of bile
Increases adrenaline secretion	Adrenal glands	
Inhibits peristalsis	Small intestine	Stimulates peristalsis
Contracts	Ileocecal valve	Relaxes
Inhibits peristalsis	Colon and rectum	Stimulates peristalsis
Contracts	Anal sphincter	Relaxes
Reduces blood flow to kidneys	Kidneys	
Contracts sphincter and relaxes muscle	Urinary bladder	Relaxes sphincter and contracts muscle
Stimulates contraction of muscles in tubes and uterus	Uterus	Relaxes uterine contractions and activates cervix
	Prostate	Contracts muscles in prostate

fight" situations. What is true in anyone with symptoms of organ dysfunction is that some organs may be in sympathetic dominance, others in parasympathetic dominance, and the rest may be behaving normally. In other words, as soon as an organ has performed whatever action was required of it to maintain the integrity of the whole, it will return to normal function.

Exhaustion of this ability can be recognized by the appearance of symptoms. These symptoms represent normal functions that are no longer occurring appropriately. They are occurring either too fast, too slow, or incompletely.

Part Two

THE GREAT FOOD ENZYME COVER-UP

What Is Food?

What Are Enzymes?

*How Enzymes Work:
The Process of Digestion*

Normal Digestion

Once you have the floor plan and have ordered the building materials, you will need workers. Enzymes have the capacity to do work. They will take the building blocks and create a house.

Chapter 3

What Is Food?

Food Is ...

... a material consisting of carbohydrates, fats, protein, and other substances (vitamins, minerals, and enzymes) that are taken or absorbed into the body of an organism in order to sustain growth, repair, and to furnish energy for all activity of the organism.

Protein is necessary for growth and development. It acts in the formation of hormones, enzymes, and antibodies. It maintains acid-alkali balance and is a source of heat and energy. Its deficiency is recognized by fatigue, loss of appetite, and edema. Prolonged deficiencies result in diarrhea and vomiting.

Carbohydrate provides energy for body functions and muscular exertions. A low-carbohydrate diet, or the inability to digest and assimilate carbohydrate, results in excessive protein breakdown to maintain blood-sugar levels. This results in loss of energy and fatigue. Prolonged deficiency causes a disturbed balance between water and the electrolytes (sodium, potassium, and chloride).

Fiber is the part of food that is not digested by the human body, such as the skin of an apple or the husk of a wheat kernel. Nevertheless, enzymes contained in raw foods will digest soluble fiber. This lowers cholesterol and fat levels in the blood. The indigestible fiber stimulates and cleans the intestinal tract, which depends on the presence of adequate fiber. A low-fiber diet has been associated with heart disease, cancer of the colon and rectum, diverticulosis, varicose veins, phlebitis, and obesity.

Fat provides energy and acts as a carrier for fat-soluble vitamins A, D, E, and K. It supplies essential fatty acids needed for growth, health, and smooth skin. A low-fat diet, or the inability to digest and assimilate fat, results in eczema and disorders of the skin and hair.

Vitamins are organic food substances found only in living things, that is, plants and animals. Each of these vitamins is present in varying quantities in specific foods, and each is absolutely necessary for proper growth and maintenance of health. With a few exceptions, the body cannot synthesize vitamins; they must be supplied in the diet or in dietary supplements.

Minerals are nutrients that exist in the body and in food in organic and inorganic combination. All tissues and internal fluids of living things contain varying quantities of minerals. Although only 4% or 5% of the human body weight is mineral matter, minerals are vital to overall mental and physical well-being. All of the minerals known to be needed by the human body must be supplied in the diet.

Enzymes are large protein molecules found in all living things. They are composed of two parts. The protein portion, or *apoenzyme*, is a long chain containing hundreds of amino acids in specific sequential arrangement. The other part, the *prosthetic group* or *coenzyme*, is usually a mineral or a vitamin, or it may contain a vitamin, or it may be a molecule that has been manufactured from a vitamin. Vitamin and mineral supplementation is wasted unless there is an adequate supply of the appropriate enzyme to utilize them.

At the present time enzymes are not considered essential because, unlike vitamins and minerals, they can be produced by the body. However, as we will see, that attitude will change in the 21st century.

Tables 6a, 6b, and 6c list all the essential ingredients found in food to sustain life. Please notice that food enzymes are not listed. Tables 6a and 6b are based on the National Research Council's Recommended Dietary Allowances. Table 6c is based on the relationships between protein, carbohydrates, and fats that were used during my years of practice.

Natural or Synthetic?

Vitamins are defined as food substances found only in living things, but they can be made synthetically. Did you know that most of the vitamins and minerals

RECOMMENDED DAILY ALLOWANCES

Vitamins

	Infants	Children	Males	Females	Units
Age	0.5–1	4–6	25–50	25–50	
Vitamin A	375	500	1,000	800	µg RE[1]
Vitamin D	10	10	5	5	µg[2]
Vitamin E	4	7	10	8	mg α-TE[3]
Vitamin K	10	20	80	65	µg
Vitamin C	35	45	60	60	mg
Folate	35	75	200	180	µg
Thiamin	0.4	0.9	1.5	1.1	mg
Riboflavin	0.5	1.1	1.7	1.3	mg
Niacin	6	12	19	15	mg NE[4]
Vitamin B_6	0.6	1.1	2.0	1.6	mg
Vitamin B_{12}	0.5	1.0	2.0	2.0	µg

[1]Retinol equivalents. 1 retinol equivalent = 1 µg retinol or 6 µg β-carotene. See *Recommended Dietary Allowances* for calculation of vitamin A activity of diets as retinol equivalents.
[2]As cholecalciferol. 10 µg cholecalciferol = 400 IU of vitamin D.
[3]α-Tocopherol equivalents. 1 mg d-α tocopherol = 1 α-TE. See *Recommended Dietary Allowances* for variation in allowances and calculation of vitamin E activity of the diet as α-tocopherol equivalents.
[4]1 NE (niacin equivalent) is equal to 1 mg of niacin or 60 mg of dietary tryptophan.

Minerals

	Infants	Children	Males	Females	Units
Age	0.5–1	4–6	25–50	25–50	
Calcium	600	800	800	800	mg
Phosphorus	500	800	800	800	mg
Iodine	50	90	150	150	µg
Iron	10	10	10	15	mg
Magnesium	60	120	350	280	mg
Selenium	15	20	70	55	µg
Zinc	5	10	15	12	mg

Age Group	Carbohydrates (grams)	Fats (grams)	Protein (grams)
Children			
0–1 year	115	28	14
1–3 years	165	38	23
4–6 years	240	58	30
7–10 years	330	80	34
Females			
11–14 years	345	80	46
15–18 years	345	78	46
19–22 years	345	79	44
23–50 years	300	66	44
Pregnant	300+	66–59	30+
Lactating	500+	66–59	20+
51 and older	277	59	44
Males			
11–14 years	330	80	45
15–18 years	330	80	56
19–22 years	360	76	56
23–50 years	390	67	56
51 and older	390	56	56

that you buy at the drugstore and health food store come from the processing of wood or petroleum? They are retrieved from the waste products formed by those industries.

Any chemist will tell you that a vitamin is a vitamin—the same molecular structure regardless of the source, natural or synthetic. Besides, the synthetic form can be concentrated into larger doses, whereas a food such as a carrot will contain only so much vitamin A or beta-carotene. Not only that, but the synthetic form has a much longer shelf life than any food source in which it may be found.

As an example, medical researchers wanted to study the role of antioxidants in the prevention of cancer. They dictated that they did not want to use carrots;

they wanted concentrated beta-carotene. There were two reasons for this:

- Too many other ingredients are found in the carrot besides the beta-carotene. When the study was finished, the researchers would not have been sure which ingredient was responsible for the results.

- The isolated beta-carotene could be concentrated into a small capsule so the subjects did not have to eat a pound of carrots every day to get a guaranteed specific amount of beta-carotene. Besides, carrots vary in the amount of beta-carotene they contain based on the soil and other conditions under which they were grown. This is all too unscientific to yield usable information.

So the studies were done with thousands of subjects, even doctors, taking a prescribed amount of beta-carotene every day for several years. The results were tabulated, and it was found that concentrated doses of beta-carotene not only did not prevent cancer, but in many subjects they increased the occurrence of cancer. Why? The answer to that question will define a major difference between the use of nutrition to maintain health and the use of medicine (drugs) to treat disease. It will also provide us with a reason that the two are not now and cannot be in the future substituted for each other.

Was the concentrated beta-carotene used in the test a food or a drug? It was a drug. Any time that you take a chemical out of a food, be it beta-carotene or some other nutrient, and concentrate it into a very large dose and feed it to human beings, you create deficiencies of the thing that you took out. When you put this into the body, there must be the other synergistic ingredients it was created within the food in order for the body to be able to utilize it. If these ingredients are not readily available in the body, the body must either tear down its own tissues to provide them, or the nutrient goes right through the body unused. Unfortunately, the body must treat it as a foreign substance and use precious energy to detoxify and otherwise process the excess. If the excess is in sufficient quantity and the body cannot prevent its accumulation, the substance will produce side effects.

Food Enzymes Are Essential Nutrients

In spite of the fact that food enzymes do not appear on the list of U.S. Recommended Daily Allowances, they are essential nutrients. All living things contain enzymes. The enzymes found in food are responsible for the biochemical reactions that bring plants to ripeness. The enzymes found in all raw foods are the workers responsible for the benefits given to vitamins and minerals, which are called coenzymes and do not have the capacity to do work.

Yet They Are Systematically Removed from the Diet

Enzymes must be removed from our food supply in order for food products to achieve extended shelf life. This process began in the early 1900s when our society began to change from a rural, agricultural base to an urban, industrial one. The canning industry needed biochemists to discover ways and means to destroy the naturally occurring enzymes found in fresh fruits and vegetables. These biochemists became the first real food enzyme experts.

Because they will digest the food in which they are contained when conditions are right for their activity, the enzymes found in raw fruits and vegetables must be removed. You can't leave the enzymes in canned food and expect to get it to market. Enzymes must be removed from food in order to prolong shelf life. Imagine what a gift it would be to a produce grocer if fresh fruits and vegetables did not contain enzymes. No more produce would be lost to spoilage.

I have books in my library in which biochemists describe the use of salicylic acid (aspirin) to destroy "the dreaded contaminant enzymes found in food." That is exactly how they described it in 1914. In the 1950s the canning industry

KEY POINT

Vitamin and mineral deficiencies result in acute symptoms, but it takes some time for enzyme deficiencies to be seen and they manifest as chronic degenerative diseases.

started holding conventions for biochemists in an effort to discover new ways to get rid of the enzymes in food.

Disinfection and the Preservation of Food

The following is an excerpt from the book *Disinfection and the Preservation of Food* published in 1903. It was written by Samuel Rideal, D.Sc., and published by the Sanitary Publishing Company of London. The book deals with removing naturally occurring enzymes from food in order to provide shelf life, and the information it contains is of interest to our present discussion. The paragraphs quoted discuss the effects of salicylic acid in removing naturally occurring enzymes from food. Salicylic acid was one of the first compounds used in that pursuit.

> Salicylic acid is by no means an innocent remedy. In fact it can be a powerful poison, as it has a disintegrating action on the blood corpuscles. <u>The salts cause albuminuria [protein being lost in the urine], hence, must be irritating to the kidneys, probably through phenol being formed.</u>(1) The acid, in strong (alcoholic) solutions, or in ointments, is so caustic that it is the basis of most of the popular cures for corns.(2) [This was true in 1903 and is still true today.] The official dose of the acid is 5 to 30 grains. Dr. Bond stated that he had taken 10 grains daily for a month without bad effect, but Dr. Brouardel has noticed daily doses of 2 grammes [sic] to produce grave symptoms of intoxication and poisoning.
>
> Kolbe (3) first drew attention to the antiseptic properties of salicylic acid. *He showed that it prevented the action of enzymes (unorganized ferments), like diastase, emulsin, and that of mustard, also gastric digestion, fermentation by*

yeast, ammoniacal fermentation of urine, and the germina-tion of seeds.[my emphasis]

In the body it undergoes a change into phenol and carbonic acid (a reverse to that in its preparation), as phenol appears in the urine. F. D. Simons, and also Chittenden, <u>found that salicylic acid and salicylate of soda greatly retard peptic digestion</u>, and H. Leffmann (4) <u>after a number of experiments concluded that 'salicylic acid in all its forms, natural, crude commercial, and refined, is distinctly antagonistic to most enzymes, especially those that convert starch.'</u>

Leffmann and Bean had shown (5) <u>that salicylic acid, 1 in 20,000 retarded the conversion of starch in the proportion of 245 to 174, or 29 percent, while 1 in 1,000 entirely prevented it, both with diastase and pancreatic ferment. A. Weber came to similar conclusions.</u>(6)

[1]*The Lancet*, Dec. 20, 1879.

[2]Whelpley, *Chem. and Drug.*, Aug. 16, 1890.

[3]*J. fur Praktisch Chem.*, 1874, vol. X, p. 89.

[4]Franklin Institute, Dec. 20, 1898.

[5]*Analyst*, 1888, 103.

[6]*Journal of the American Chemical Society*, 1892, p. 4.

You can see that the blood-thinning qualities of salicylic acid (aspirin) were well known in 1903. The second sentence in the preceding article stated that it had "a disintegrating action on the blood corpuscles (cells)." In today's parlance we would say that it thins the blood. Equally alarming is the "irritating effect it has on the kidneys, probably through phenol being formed." Phenol is toxic to the body and irritates the tissues as it circulates through the blood. Unfortunately, the liver, which is responsible for detoxifying toxins, cannot detoxify phenol. The relevance of this information is that phenol is also formed in the lower bowel when inadequately digested food putrefies. These chemicals are so irritating to the bowel that the body is obliged to absorb them across the gut wall into the blood so they can eventually be eliminated through the kidney.

Obviously, these effects of salicylic acid are to be avoided as much as possible. But please note that back when that book was written, these effects of salicylic acid were deemed desirable when used to destroy enzymes and preserve food because "it prevented the action of enzymes, like diastase [amylase] … as well as gastric digestion … Chittenden found that salicylic acid and salicylate of soda retard peptic (stomach) digestion … Salicylic acid in all its forms … is distinctly antagonistic to most enzymes …"

The removal of enzymes to preserve food and extend shelf life has come a long way since 1903. It has developed into a very advanced science and no longer uses salicylic acid to preserve food. Many new technologies have been advanced to destroy the enzymes found in food. One of the newest areas of advancement is in growing hybrid foods such as tomatoes that will have a reduced amount of naturally occurring enzymes. This will allow greater shelf life in the produce section of the grocery store. It will also place a greater strain on individuals who consume such foods to digest them.

How Are Enzymes Replaced in the Diet?

While we are very careful to replace the vitamins and minerals lost in the processing of food, we do not replace the more important food enzymes. Unless supplemental food enzymes are taken, our diets will continue to be deficient in these essential nutrients.

What Are the Symptoms of Enzyme Deficiency?

Coenzyme deficiencies produce acute symptoms, while food enzyme deficiencies are more insidious, producing chronic degenerative changes.

Enzymes Digest Food Before Stomach Acid Is Secreted

It takes 30 to 60 minutes for the stomach to concentrate enough stomach acid to reduce the resting pH of the stomach from between 5.0 and 6.0 down to 3.0. Studies indicate that most geriatric patients are unable to do this at all! For this segment of the population, food enzyme predigestion would appear to be mandatory!

Enzymes Deliver Nutrients Past an Incompetent Digestive System

Food enzymes work in a pH range of 3.0 to 9.0 and are able to predigest food in the stomach before they are inactivated. This allows 60% of starch, 30% of protein, and 10% of fat to be predigested, thus supplying nutrients that otherwise would not be available to nourish the body.

Animal Enzymes Do Not Predigest Food

Pancreatic enzymes work only in an alkaline environment of approximately 7.2 to 9.0 pH—in other words, in the duodenum only. They cannot function in the stomach and thus do not spare the body the necessity of providing all of the enzymes needed to digest the food.

FIGURE 3

ENZYMES EXPLAINED

The terms food enzymes, plant enzymes, and animal enzymes are ambiguous and confusing. I hope the following diagram will be helpful.

All living things contain enzymes, animal and plant.
Therefore, we use the term food enzyme inclusively.

FOOD ENZYMES
Since food enzymes must be removed from the diet,
they should be replaced.

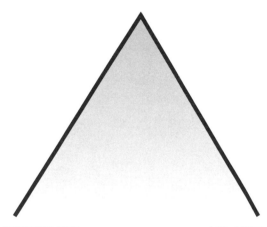

ANIMAL ENZYMES
Supplemental animal enzymes are derived from the pancreas of beef and pork. They are sold under the generic name of pancreatin. They do not spare the body's own digestive resources. In other words, the body still must contribute a large majority of its own enzymes to digest food in the early stages of digestion.

PLANT ENZYMES
Plant enzymes work in a broader pH range than animal enzymes and therefore predigest food in the stomach.

Chapter 4

What Are Enzymes?

The word *enzyme* comes from the Greek *en*, "in," and *zyme*, "leaven." Enzymes are proteins secreted by cells that act as catalysts to induce chemical changes in other substances without undergoing change themselves. Catalysts are used in very small amounts (called catalytic amounts) compared with the amounts of reactants that are consumed in the reaction. Dr. Howell took exception to describing enzymes as catalysts. He believed that this was accurate for inorganic catalysts, but he presented evidence to show that enzymes eventually do wear out as they are discarded by the body.

Composition of Enzymes

Essentially an enzyme is composed of two parts: the protein portion and the prosthetic group. The protein portion of the enzyme is called the apoenzyme. The apoenzyme requires a prosthetic group (cofactor or coenzyme) to become a functioning enzyme.

The Apo-Portion (Amino Acids and Peptide Bonds)

A protein is a very large molecule made up by bonding together many much smaller molecules called *amino acids*. Proteins are composed of many amino acids all held together by peptide bonds.

A number of amino acids are essential in the diet, that is, they are needed for building body proteins, and they cannot be synthesized by the body from other available materials. There are 21 of these amino acids, which occur commonly in enzymes.

Proteins are built up by combining the amino group ($-NH_2$) of one amino acid with the carboxyl group ($-COOH$) of another, and eliminating a molecule of water in the process.

Notice that the amino group is attached to the carbon atom adjacent to the carboxyl group in all cases. Amino acids having this makeup are called *alpha amino acids*.

FIGURE 4

FORMATION OF THE PEPTIDE BOND

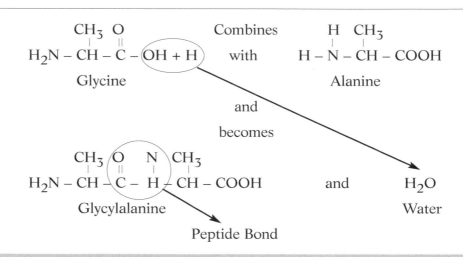

The process is opposite that of digestion as we will discuss in Chapter 5. Proteins are formed by forming peptide bonds between amino acids and eliminating water. Proteins are digested when enzymes break the peptide bonds and add a molecule of water to the broken ends.

In splitting out water from the $-NH_2$ and $-COOH$ groups, a bond called a *peptide bond* ($-CONH-$) is formed.

Ribonuclease, the first enzyme to be synthesized in the laboratory, contains 124 amino acid residues (amino acids with a water molecule split out in forming the bond). This makes it a very small protein. Some protein molecules contain several hundred amino acid residues. Ribonuclease has a molecular weight of around 14,000, and many protein molecular weights are in the hundreds of thousands.

Where Do Supplemental Enzymes Come From?

All living organisms depend on enzymes, so all living organisms may be considered sources of enzymes. This includes microorganisms, plants, animals, and humans. Microorganisms are an excellent source of enzymes not only because they can be grown in small quantities or in industrial-scale quantities, but also because many of these enzymes are secreted by the microorganism into the medium in which they grow. This greatly facilitates the isolation of the enzyme.

Plant enzymes are usually obtained by grinding the plant material and extracting the ground vegetable material to separate the soluble protein from the insoluble matter. Animal enzymes are similarly obtained. Usually it is determined which organ is particularly rich in the desired enzyme, and that particular organ is used as the source of the enzyme.

Purification of Enzymes

It can be seen that the enzymes are derived from complex materials and that they will be mixed with many types of soluble material. This necessitates complex purification procedures. The degree of purification will depend, to a large extent, on the intended use of the enzyme. For example, an enzyme used to treat cowhide in leather manufacturing (such as papain) does not need the degree of purification required by enzyme preparations for human consumption.

The development of purification techniques has been largely responsible for the tremendous gains in knowledge in the enzyme field in the last 30 years. Biochemists, studying enzymes in dilute aqueous solution, have had progressively purer catalysts for their studies. This greater purification has also led to more practical uses. More enzymes are being made available that are pure enough for use in treating sick humans and animals, and the results are quite impressive.

Enzymes and Coenzymes

Minerals are nutrients that also exist in the body and in food. They are simple elements found not just in living things, unlike vitamins. They are nevertheless essential for life. Only about 4% or 5% of the human body weight is mineral. Yet as with vitamins, we often supplement them in massive quantities—zinc for prostate problems, calcium for osteoporosis, and chromium for diabetes. Have you ever thought about what happens to these minerals when you put them into a body? How does the body use them?

Vitamins and minerals are called coenzymes. What does that mean? As mentioned before, vitamins, minerals, protein, carbohydrates, and fats are building blocks. If you took high school physics, you remember that the definition of energy is the capacity to do work. Protein, carbohydrates, fats, fibers, vitamins, and minerals do not have the capacity to do work. They are certainly needed to fuel the body that will do the work, but they do not have the capacity to do work any more than concrete blocks have the capacity to build a building.

The Coenzyme Factor

The apoenzyme requires a prosthetic group (cofactor or coenzyme) to become a functioning enzyme. Cofactors or coenzymes are relatively small organic compounds that are required for the activity of many enzymes. Many cofactors are converted by the enzyme reaction to a form that no longer has the power of assisting the enzyme in catalyzing the reaction. Therefore, the reaction mixture must contain one molecule of cofactor for every molecule of substrate to be converted. Many of the vitamins that are essential to human health function as part of enzyme cofactors.

Many enzymes require the presence of metal ions for activity. A metal ion is an atom of metal that has lost one or more electrons and therefore has a positive electrical charge. The metal ions most frequently used as enzyme activators are magnesium, manganese, zinc, potassium, iron, copper, calcium, cobalt, and molybdenum.

Some metal ions fall into the category of inhibitors. Inhibitors are atoms, ions, or molecules that retard or terminate enzymatic activity. They are classified as *competitive* and *noncompetitive*.

- A *competitive inhibitor* is a substance that combines with the active site of the enzyme, preventing the substrate from having access to that site and thus preventing reaction. The active site of the enzyme is the part of the enzyme molecule that must combine with the substrate before any conversion is affected. A competitive inhibitor blocks this site.

- A *noncompetitive inhibitor* combines with the enzyme at a location other than the active site. It does not affect the bonding of the substrate to the enzyme, but nevertheless it retards the conversion of the substrate to product.

Classification and Naming of Enzymes

Enzymes, with the exception of those discovered long ago (such as pepsin), are generally named by adding *-ase* to:

- the name of the substrate on which the enzyme acts (such as glucosidase),
- the substance activated (such as hydrogenase), or
- the type of reaction involved.

All known enzymes have been classified into six fundamental groups based on the types of reactions they catalyze. These groups are hydrolases, lysases, oxidoreductases, transferases, isomerases, and ligases.

Hydrolases

Hydrolyases digest food. Hydrolytic enzyme reactions are ones in which chemical bonds are broken with the addition of water. Hydrogen (H+) is added to one of the resulting fragment molecules, and a hydroxyl radical (OH–) is added to the other. This is the reversal of the process we saw in peptide bond formation, where water was split out when the bond was formed.

Lysases

Lysases catalyze the decomposition of a molecule into two fragments. They catalyze the cleavage of chemical bonds without adding water to the resulting fragments.

Oxidoreductases

Oxidoreductases are involved with transfer of electrons from one molecule to another. No molecule can be oxidized without the simultaneous reduction of another molecule. Coenzymes are usually active in conjunction with oxidoreductases.

Transferases

Transferases catalyze the transfer of a chemical group from one molecule to another. They transfer chemical groups such as $-CH_3$, $-PO_4$, and $-NH_2$ from donors to acceptors.

Isomerases

Isomerases change the arrangement of molecules within a substrate. They convert a molecule into the mirror-image of the original. A conversion of a D-form (dextrorotatory) to an L-form (levorotatory) would be an isomerization.

Ligases (synthetases)

Ligases catalyze the formation of a chemical bond between two molecules that enable energy-rich phosphate compounds to be broken down. An excellent example is the conversion of glucose into energy inside the cells of the body.

Quite frequently in scientific papers, authors will mention the systematic name once and thereafter use a more familiar, "trivial" name. Trivial names bear no relationship to either the substrate or the reaction hydrolyzed. These include papain, trypsin, lysozyme, and many more. For example, we shall have occasion to discuss an enzyme systematically named beta-D-galactoside galactohydrolase. We shall use its trivial name, lactase, or a somewhat more elegant name, beta-galactosidase.

When Do Enzymes Work?

One of the most important properties of enzymes is their specificity—that is, their catalytic effect is limited to a single kind of molecule, or to a relatively small group of related molecules. This is a very important point to remember, especially when someone tells you enzymes do not make it through the stomach without being destroyed. There is nothing in the stomach to digest (hydrolyze) an enzyme—certainly not stomach acid, which only activates the enzyme pepsin! The effectiveness of enzymes depends on the environment in which the enzyme acts. Three important determinants are the temperature, the concentration of substrate, and the acidity or alkalinity of the solution.

Temperature

In general, the rate of a chemical reaction is approximately doubled when the temperature is increased by 10 degrees Centigrade. This applies to enzyme-catalyzed reactions with the limitation that excessively high temperatures denature the protein, destroying the activity. Thus, the rates of enzyme reactions increase with the temperature, until a point is reached at which the protein becomes unstable. The temperature at which the enzyme-catalyzed reaction goes most rapidly is called the optimum temperature for that enzyme.

Substrate Concentration

The rate of reaction depends on the substrate concentrate. The substrate concentration is much greater than that of the enzyme. By increasing the concentration of substrate, the reaction rate increases until a point is reached at which the enzyme is saturated with substrate and cannot handle more. At that point, the addition of more substrate does not accelerate the reaction.

pH Environment

Enzymes work in very specific pH ranges. For example, plant enzymes work roughly in a pH range of 3.0 to 9.0. Animal (pancreatic) enzymes work only in alkaline pH ranges, 7.0 to 9.0.

General Types of Hydrolytic Digestive Enzymes

Protease or Peptidase

- High levels found in meats (before cooked), eggs, milk, and natural cheese.
- High levels found in soy, wheat, barley, bulgur, wild rice, and peanuts.

Proteases (protein-ase) hydrolyze proteins. They split large polypeptides by breaking and adding a molecule of water to the broken ends and forming smaller polypeptides. They have been shown to be absorbed in substantial quantities into the blood where they bind to serum proteins and are utilized by tissues involved with immune function. Plant or food proteases have been proven to be effective in dissolving blood clots (fibrinolysis) and restoring normal blood flow.

Lipase

- Very high levels found in avocados and olives.
- High levels found in nuts and seeds.
- High levels found in some fruits like bananas, cherries, figs, and grapes.

Lipase (lipid-ase) hydrolyzes fats into monoglycerides and fatty acids. The role of the enzyme lipase is being studied extensively in treatment of chronic pancreatitis, pancreatic cancer, malabsorption abnormalities, myocardial infarction, and cholesterol and triglyceride studies.

Catalase

Catalase is an oxidoreductase enzyme. It hydrolyzes hydrogen peroxide (H_2O_2) to water and oxygen. Peroxide is formed during energy production inside every cell and is toxic to the human body. Therefore, it must be destroyed for normal cellular function to continue. Anaerobic and lactic acid bacteria do not produce peroxide and therefore do not contain catalase. Other oxidoreductases in the body include superoxide dismutase and glutathione peroxidase.

Amylases

The most complicated food substrates for the body to digest are carbohydrates. There are many steps involved in its hydrolysis. First, the body must strip any protective cellulose layer from vegetables. It must do this without having an enzyme capable of digesting cellulose. Then the body uses salivary and pancreatic amylase to digest starch into simple sugars. Finally, it digests the simple sugars to glucose. However, some complex sugars, like those found in beans, for example, have always defied digestion for some people.

Cellulase

- High levels found in vegetables and grains, especially wheat and millet.
- High levels found in fibrous fruits like apples, papaya, pears, and some melons.

Cellulase is not found in humans. It occurs in various bacteria, fungi, plants, and lower animals. It digests only soluble fiber by cleaving internal glycosidic linkages, such as occur in cellulose. Plant fiber (cellulose) cannot be broken down in the human body.

Yet cellulose covers most vegetables as a thin, filmy covering. This dilemma must be resolved; otherwise a great deal of intestinal gas may be formed when people eat raw vegetables. Fortunately, cooking destroys this covering; thus, the nutrition textbooks hail cooking as a great aid to digestion. Unfortunately, cooking denatures the enzymes contained in the vegetable. The books never mention that chewing also removes the cellulose layer without destroying the enzyme content. People who claim they cannot eat raw food because of the discomfort from intestinal gas are really saying they do not chew raw food thoroughly.

If the body could digest cellulose, we would find it to be a very satisfactory food source! Cellulose is similar to disaccharides in that it contains a molecule of glucose bonded to another molecule. You may be surprised to learn that the other molecule is a short-chain fatty acid.

The addition of cellulase to food supplements will assist in the digestion of raw vegetables and still allow their fiber to assist in cleansing the bowel. The short-chain fatty acids will be a source of nutrition to the body and assist in fat metabolism without elevating cholesterol levels.

More important to some is the fact that cellulase also digests that group of microorganisms known as fungi/yeasts. There are some 2,500 such organisms known, of which *Candida albicans* is only one.

Amylase

• High levels found in grains and fruits.

Amylase digests starches or complex carbohydrates. Digestion of carbohydrates begins in the mouth as food is chewed and the salivary glands secrete amylase. Supplementary plant amylase allows digestion of carbohydrates to continue in the stomach for about 45 minutes, until stomach acid is produced. The pancreas also secretes amylase, which acts in the small intestine.

Pancreatic and salivary amylase are identical except for the pH range in which they work (pancreatic 7.2 to 9.0 and salivary 6.3 to 9.0). These amylases, produced in the body, require the presence of a halogen ion such as chlorine, bromine, iodine, or fluorine to work. This is not true of plant amylases.

Glucoamylase

Glucoamylase is a glycosidase. This is an enzyme with the ability to do what the human gastrointestinal tract cannot: break down raffinose and stachyose (glucosides) into simple sugars that the body can digest. Each enzyme within this general class is specific for a particular monosaccharide ring structure. However, glucoamylase is highly specific in its recognition of glucose, and its only substrate is starch.

Raffinose and stachyose are abundant in beans, peas, nuts, and seeds. These sugars are also found in lesser amounts in grains (oats and oat bran, whole-wheat and white flour, amaranth, barley, and millet) and in many vegetables such as beets, broccoli, pumpkin, and cabbage.

Oddly, of the hundreds of enzymes in the human body, none can digest raffinose or stachyose. Too large to pass through the intestinal wall, these undigested sugars remain in the gut, where bacteria begin to feed on them. As the bacteria digest them, the sugars ferment, producing substantial amounts of gases. Through the enzymatic cleavage of the glucosidic bonds in these molecules, the intestinal discomfort and flatulence associated with beans can be abated.

Alpha-Galactosidase

This enzyme is normally present in cellular lysosomes (discussed in Chapter 8) but is absent from the digestive juice. Without this enzyme, foods rich in galactose bonds (e.g., beans) provide undigested nutrients to bacteria in the colon, which digest them and produce gas.

Phytase

An enzyme that catalyzes the breakdown of phytic acid (inositol hexaphosphoric acid), which is found in grains, seeds, rhizomes, and other food.

Pectinase

An enzyme that breaks down pectin, a noncellulose polysaccharide commonly found in fruits and vegetables. The purpose of this is not to produce nutrients for absorption, but to increase the solubility of the fiber to increase its ability to bind water and toxins.

Disaccharidases

Disaccharidases digest the simple sugars that are found in dairy products (lactose), grains (maltose), and white sugar and flour (sucrose).

Lactase

Reduced or absent lactase activity results in lactose intolerance. B-galactosidase (lactase) occurs in the brush border membrane of the intestinal mucosa. This enzyme catalyzes the hydrolytic cleavage of lactose to galactose and glucose and also cleaves terminal nonreducing galactose residues from some B-glycosides.

Malt-Diastase or Maltase

An enzyme closely related to amylase, malt-diastase attacks carbohydrates at their ends, cleaving off disaccharides that are then broken down by enzymes bound to the brush border of epithelial cells lining the small intestine. The final products, simple sugars, are readily absorbed into circulation.

Invertase

An enzyme that breaks sucrose down into glucose and fructose, both of which are readily absorbed into the blood. This enzymatic function is normally performed by enzymes secreted by the epithelial cells lining the small intestine.

These simple sugar–digesting disaccharidases are important to stimulate the secretion of insulin, a hormone that promotes the efficient storage and utilization of these fuel molecules by controlling and regulating various intracellular biosynthetic pathways.

ENZYMES: *The Key to Health*

Chapter 5

How Enzymes Work:
The Process of Digestion

What Is Digestion?

The digestion of food is the process of hydrolysis, or the addition of water. The function of digestion is to reduce food particles from their combined form, in long-chained molecules, to their smaller basic components. This is necessary so they may pass across the gut wall.

FIGURE 5

CARBOHYDRATE DIGESTION

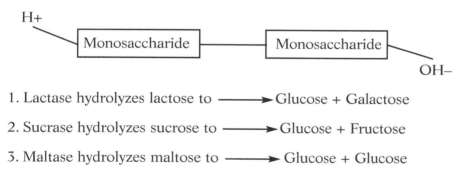

1. Lactase hydrolyzes lactose to ⟶ Glucose + Galactose

2. Sucrase hydrolyzes sucrose to ⟶ Glucose + Fructose

3. Maltase hydrolyzes maltose to ⟶ Glucose + Glucose

Carbohydrate digestion (Figure 5) is very similar to protein digestion but varies in that amylase does not reduce the chains down to individual monosaccharides. Rather, it reduces them to the disaccharide level. In other words, it only breaks every other bond in the chain, thus leaving two monosaccharides bound together that cannot be utilized by the body if absorbed across the gut wall. The disaccharides are lactose, sucrose, and maltose.

These molecules must be digested further to monosaccharides by the specific action of enzymes that are made by the brush border of the small intestine.

FIGURE 6

PROTEIN DIGESTION

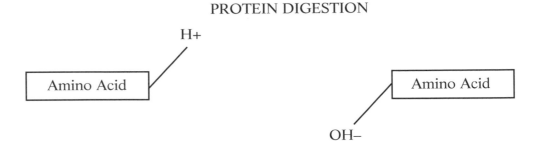

Protein digestion (Figure 6) requires breaking long polypeptide chains into amino acids. This is accomplished when a proteolytic enzyme breaks the bond between amino acids and attaches a H+ ion to one end and a OH– radical to the other.

FIGURE 7

FAT DIGESTION

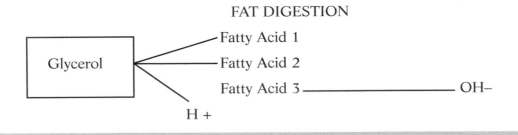

Fat digestion (Figure 7) requires a similar reaction, but its structure differs from that of protein and carbohydrate in that three fatty acids attach to a molecule of glycerol, and the action of bile is necessary to expose the bonds to allow lipase to hydrolyze them. There are no enzymes in bile itself.

The Secret to Improving Digestion

The digestion of food is largely taken for granted by just about everyone. Most books on the science of nutrition describe the normal process of digestion but do not attempt to unravel the secrets of poor digestion.

ENZYMES: *The Key to Health*

More than $1 billion is spent annually on drugs to relieve heartburn, excess acid, bloating, and other symptoms of indigestion. These products are designed to give temporary relief or to cover up these symptoms. They do nothing to improve people's ability to digest food on their own. Fortunately, there is a little-known secret about improving digestion that has been around for at least 50 years; only now is it receiving attention, but not nearly as much as it deserves.

Simply stated, the secret is that each raw, uncooked fruit, vegetable, or meat contains enzymes that will digest the food in which they are contained.

The problem is that these enzymes are destroyed during cooking, canning, and other methods of food processing. In fact, cooking is stressed as a major digestive aid in nutrition textbooks. While cooking does function to make it easier to digest some foods, like raw vegetables, it also destroys enzymes contained in the food that would improve digestion even more than cooking. Temperatures above 118 degrees Fahrenheit destroy the enzymes found in our food. Also, many foods we eat today are man-made and do not contain enzymes.

Enzymes run virtually all of the biochemical processes in living things. In the case of plants, these enzymes bring the plant to maturity or ripeness. And they will digest that plant when they are properly activated. All that is required is for the enzymes to be released by the chewing or cutting of the plant. All enzymes require the presence of water, the proper temperature, and correct pH range in order to work, and those conditions are present in the mouth and saliva.

Digestive Problems Caused by Enzyme Deficiencies

When these indigenous food enzymes are destroyed, your body must assume the entire burden of digesting the food. Scientists are gradually becoming aware that the organs that produce our digestive enzymes are not large enough to produce all the enzymes needed to digest the average American diet.

Major problems can arise when the body is unable to digest and assimilate food that has been put into it. Nature's plan calls for the enzymes found in raw food to help with digestion.

If food enzymes do some of the work, the body is not burdened with eliminating an accumulation of food it cannot assimilate. Food allergies, gas and bloating, heartburn, constipation, or diarrhea are only minor problems that can result. Studies are gradually revealing that the resulting metabolic problems may be the direct cause of many chronic degenerative diseases.

Edward Howell, M.D.

In the 1930s Dr. Howell postulated his food enzyme concept and presented it in an article entitled "Are Food Enzymes Important in Digestion and Metabolism?" He later expanded on his ideas and presented them in his first book, *Food Enzymes for Health and Longevity*, published in 1946. He presented several key points that seem to have been missed in the evolution of nutritional thought.

Dr. Howell's Food Enzyme Paradigm

1. There are three broad classifications of enzymes.
    a. Those that occur in our food.
    b. Those that are made in the body for digestion of food.
    c. Metabolic enzymes made to run the biochemical reactions occurring in the body.
2. The presence of a predigestive stomach in humans, just as in other animals and mammals.
3. The concept of pancreatic hypertrophy. Howell was the first to realize that the pancreas enlarges when it is required to produce all of the enzymes needed to digest the diet.

Howell believed that when we eat cooked enzyme-deficient foods, the body is forced to produce enzymes needed for digestion. He proved by autopsy

ENZYMES: *The Key to Health*

weights that cattle fed a diet poor in naturally occurring food enzymes had a larger pancreas per percentage of body weight than cattle who were allowed to graze in pastures. He believed this proved that the pancreas was never intended to make all the enzymes needed to digest our entire dietary intake. He believed the body had to borrow enzymes from other tissues and organs to complete digestion. Today we know that our immune system must make up for any pancreatic insufficiency. Howell believed that this stealing of enzymes from other parts of the body sets up a competition for enzymes among the various organ systems and tissues of the body, and the resulting metabolic dislocations may be the direct cause of chronic degenerative diseases.

The Importance of Predigestion

Digestion begins in the mouth. The salivary glands secrete some enzymes that begin working right away, if the food is chewed thoroughly. If the food contains its own enzymes, they also begin to work. If it is cooked or processed, only the enzymes in the saliva are available to begin digestion.

Digestion continues after food is swallowed. This food passes into the upper or cardiac part of the stomach, where it sits for up to one hour waiting for the body to produce enough acidity to activate its enzymes for protein digestion. The pH of the stomach during this time is only slightly acid, just right for the food enzymes to work, but not acidic enough for the body's own digestion to begin. It is during this time that the salivary enzymes and food enzymes do their best work digesting protein, carbohydrates, fats, and fibers in the food.

Studies conducted at the University of Illinois and Northwestern University have shown that as much as 45% of ingested carbohydrates can be digested after the first 15 minutes in the stomach when only salivary amylase is at work. Imagine the amount of digestion that can occur with the addition of supplemental enzymes.

In response to the stretching of the stomach wall when food is eaten, the stomach begins to concentrate acid to lower its pH to below 3.0. This usually takes 30 to 60 minutes, but recent studies indicate it takes much longer for older patients to do this, if they can do it at all. Again, it is not true that hydrochloric acid digests food. Remember, hydrochloric acid only provides the acid environment to begin the body's digestive process. It takes 30 to 60 minutes for this acid to be produced, and in many older adults it can hardly be produced at all.

As the stomach gradually becomes more acid, any food predigested (liquefied) by food enzymes moves down the stomach toward its opening into the intestine. When this food (now called chyme) passes into the intestine, the body sends a hormonal signal to the pancreas, informing it as to the amount of protein, carbohydrate, and fat still left to be digested. The pancreas now prepares the exact amount of its own digestive enzymes needed to finish the job.

These enzymes work only when the pH of the intestine is slightly on the alkaline side; remember that the food coming from the stomach has been acidified. So the body must also provide enough alkalinity for the enzymes to work.

You can see how difficult the process of normal digestion is, and the many places its normal processes can malfunction. It also can be seen that predigestion, accomplished by food enzymes that work in both an acid and alkaline environment, can dramatically improve any digestive problem.

Pancreatic Enzymes Do Not Predigest Food

Pancreatic enzymes can be obtained from pigs and cattle. They are sold as food supplements and are even given by prescription, which often contains a pain reliever. As mentioned previously, pancreatic enzymes work only in the alkaline environment of the small intestine. This condition is not found in the stomach; therefore, pancreatic enzymes do not predigest food as food enzymes do. They certainly do digest food in the small intestine (provided that the blood can furnish adequate amounts of alkalinity); however, they cannot spare the body the necessity of providing 100% of the enzymes needed to digest food.

What Are the Consequences of Poor Digestion?

When any tissue of the body cannot secrete enough substance it is responsible for providing, it hypertrophies (enlarges)—that is, it swells and gets larger trying to provide more tissue to make more secretion. This is a well-known phenomenon in thyroid disorders. It is seen in some heart conditions when the heart enlarges in a vain attempt to meet its obligations. It is also commonly seen in prostate glands and ovaries. It also occurs in the pancreas when it consistently cannot supply all of the enzymes needed to digest the diet.

The Body's Own Digestive Enzymes

There are about 24 digestive enzymes that are manufactured and secreted by the body. The most important are the following:

Saliva

Amylase

Stomach

Pepsin

Lipase

Cathepsin

Small Intestine

Enterokinase

Dipeptidase

Saccharidases

Three nucleases

Maltase

Sucrase

Lactase

Pancreas

Trypsin	Cholinestease
Chymotrypsin	Phospholipase
Ribonuclease	Pancreatic amylase
Deoxyribonuclease	Pancreatic lipase
Carboxypolypeptidase	

Following are the major contents of bile; however, please be aware that there are no enzymes in bile.

Liver (Bile)

Bile pigments

Bile salts

Cholesterol

Various others

ENZYMES: *The Key to Health*

Chapter 6

Normal Digestion

Digestion Begins in the Mouth ...

Step 1. Chewing breaks up large food particles into smaller particles. The importance of this step in the digestive process is often overlooked. Not only is it needed so that food can be swallowed without choking, it is also necessary to expose as much surface area as possible on the particles so enzymes can begin digestion.

Step 2. The salivary glands secrete mucus into the mouth, which moistens and lubricates the food particles prior to swallowing. Saliva also contains enzymes.

- Amylase is secreted from the parotid glands and breaks down carbohydrates into smaller molecules.
- Protease is secreted from the submandibular glands and begins protein digestion.
- Lipase is secreted from the sublingual (under the tongue) glands to initiate fat digestion.

What is often forgotten is that the enzymes contained in the food being eaten (if any are present) also begin working. One of these, cellulase, is not made by the human body. It digests any soluble fiber present. This is of critical importance since those vegetables that contain cellulase are covered with a thin coating of cellulose. If that cellulose is not removed by cooking, then it must be chewed off because human enzymes cannot penetrate that protective layer. If it is not removed, the person then develops gas when eating raw food.

Step 3. A third function of saliva is to dissolve some of the molecules in the food to activate the chemical receptors in the mouth, giving rise to the sensation of taste.

... and Continues in the Predigestive Stomach

Swallowing moves the food into the pharynx and esophagus, which contribute nothing to digestion but simply provide the pathway by which ingested food and drink reach the stomach.

The stomach is flat when it is empty. When food enters in the upper part of the stomach (cardiac portion), it begins to stretch the stomach. As more food arrives, it stretches the stomach wall further. This stretching signals the body to begin forming hydrochloric acid. It will take about 45 minutes for this acid to be formed and begin its digestive work. This time period is normal for young, healthy adults. Most people past the age of 40 will require much longer periods. Most of those applying for Social Security cannot concentrate enough stomach acid to adequately initiate protein digestion.

It is this physiological time period, a minimum of 30 to 60 minutes, that Dr. Howell called the predigestive stomach. All animals and mammals have an extra stomach where the enzymes contained in the food eaten begin digesting food. Humans do not have an extra chamber for this process to occur, but we do have this time period when the food is being stored waiting for hydrochloric acid to be formed.

During this period, the salivary enzymes and any enzymes contained in the food ingested will continue the process begun in the mouth. Studies indicate that at least 40% and up to 80% of the starch (complex carbohydrate) can be digested within 15 minutes! Averages for each food component are 60% starch, 30% protein, and 10% fat. All this before the stomach begins its own digestive processes.

FIGURE 8

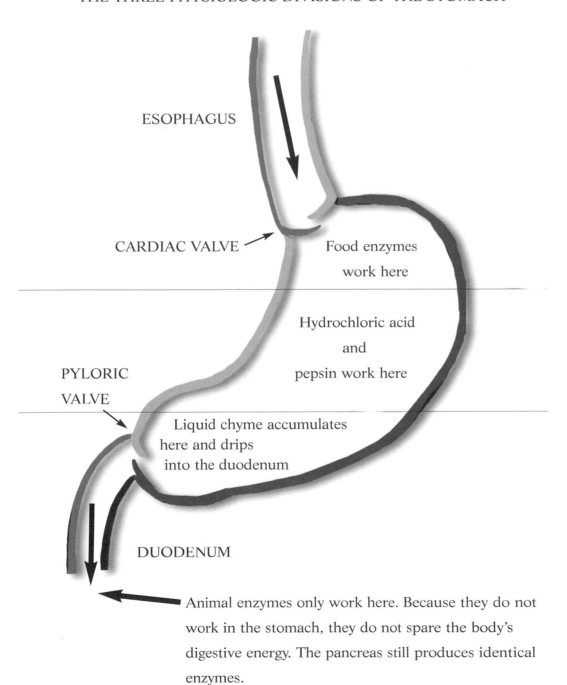

THE THREE PHYSIOLOGIC DIVISIONS OF THE STOMACH

ESOPHAGUS

CARDIAC VALVE

Food enzymes
work here

Hydrochloric acid
and
pepsin work here

PYLORIC

VALVE

Liquid chyme accumulates
here and drips
into the duodenum

DUODENUM

Animal enzymes only work here. Because they do not
work in the stomach, they do not spare the body's
digestive energy. The pancreas still produces identical
enzymes.

FIGURE 9

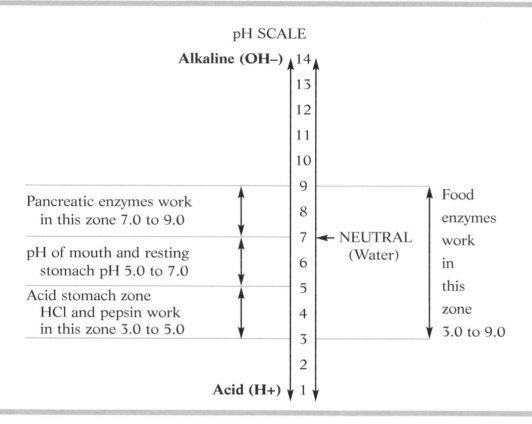

pH SCALE

Alkaline (OH–) 14 13 12 11 10 9 8 7 ← NEUTRAL (Water) 6 5 4 3 2 Acid (H+) 1

Pancreatic enzymes work in this zone 7.0 to 9.0

pH of mouth and resting stomach pH 5.0 to 7.0

Acid stomach zone HCl and pepsin work in this zone 3.0 to 5.0

Food enzymes work in this zone 3.0 to 9.0

The Acidity of the Stomach

The chief cells in the middle portion of the stomach (fundus) secrete a protein-digesting enzyme known as *pepsinogen*. Pepsinogen is an inactive enzyme. It requires the presence of hydrochloric acid in order to begin digesting protein. The major role of HCl is to activate pepsinogen, which now becomes known as pepsin. It is pepsin that splits protein into small peptide fragments.

You will often hear those not familiar with digestion say that hydrochloric acid digests enzymes, because they are proteins, as soon as they move into the stomach. As you can see, hydrochloric acid does not digest food (only enzymes can do that); rather, it activates your protein-digesting enzymes.

A second function of gastric acid is to kill most of the bacteria that enter along with food. This process is not 100% effective, and some bacteria survive to take up residence and multiply in the intestinal tract, particularly the large intestine.

Stomach acid is actually not made by the cells of the stomach as pepsinogen is. The ingredients for hydrochloric acid, namely hydrogen (H+) and chloride (Cl–), are donated from the blood. They pass through the parietal cells and are combined only inside the stomach. This is an important point because stomach acid could easily destroy the wall of the stomach if it were not protected by a thick layer of mucus. Like the salivary glands in the mouth, mucus is secreted by cells in the stomach to protect it. When a person needs antacids for indigestion, it is not because there is too much acid; it is because the stomach cannot produce a good-quality mucus to protect itself.

FIGURE 10

HOW STOMACH ACID IS MADE

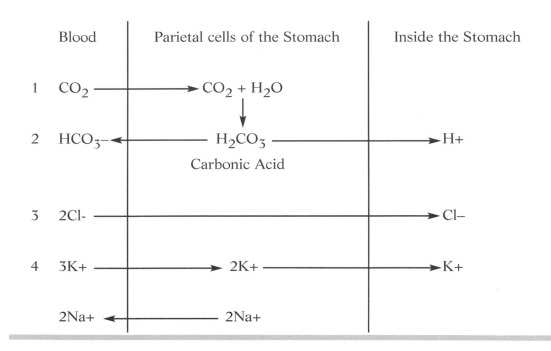

Figure 10 illustrates how stomach acid is made. It shows how the ingredients for stomach acid actually come out of the blood!

Step 1. The body takes carbon dioxide (CO_2) from the blood and puts it into the parietal cells of the stomach, where it combines with water to form carbonic acid (H_2CO_3).

Step 2. Carbonic acid is divided by an enzyme called carbonic anhydrase into its two parts: hydrogen ($H+$), which is acid in solution, and bicarbonate (HCO_3-), which is alkaline. The hydrogen is put inside the stomach, and the bicarbonate is returned to the blood.

Step 3. Chloride ($Cl-$) is transported from the blood directly into the stomach, where it combines with hydrogen to form stomach acid.

Step 4. The body also exchanges potassium ($K+$) for sodium ($Na+$) as part of its placing chloride into the stomach.

FIGURE 11

HOW THE DUODENUM BECOMES ALKALINE

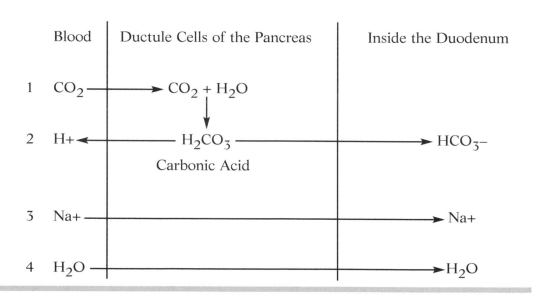

Figure 11 illustrates how alkalinity is placed into the duodenum to activate the pancreatic enzymes. It shows how the ingredients are taken out of the blood!

Step 1. The body takes carbon dioxide (CO_2) from the blood and puts it into the ductule cells of the pancreas, where it combines with water to form carbonic acid (H_2CO_3).

Step 2. Carbonic acid is divided by an enzyme called carbonic anhydrase into its two parts: hydrogen ($H+$), which is acid in solution, and bicarbonate (HCO_3-), which is alkaline. The hydrogen is placed back into the blood, and the bicarbonate is placed into the duodenum.

Step 3. Sodium ($Na+$) is transported from the blood directly into the duodenum, where it combines with the bicarbonate to alkalize the intestine.

Step 4. The body also carries water directly from the blood into the duodenum.

The Alkalinity of the Duodenum

By the time partially digested food reaches the bottom of the stomach (pylorus), it is a liquid and very acidic. As it is being formed, it drips through the pyloric valve onto the wall of the upper portion of the small intestine (duodenum). This partially digested food (chyme) activates the formation of two hormones, secretin and cholecystokinin. These hormones are carried by the bloodstream to the pancreas and biliary portion of the liver.

The exocrine portion of the pancreas secretes digestive enzymes specific for each of the three classes of food components—protein, carbohydrate, and fats. These enzymes enter the small intestine through a duct leading from the pancreas to the duodenum. The importance of secretin and cholecystokinin is that they have carried information regarding the amount of undigested protein, carbohydrate, and fat that is in the chyme. In other words, the more predigestive work that was done in the stomach, the less work the pancreas is required to do. This is of incredible importance to us as more and more enzyme-deficient food is included in our diets.

Another important point is that these pancreatic enzymes can be active only in an alkaline environment, unlike stomach pepsin, which is active only in an acid environment. The high acidity of the chyme coming from the stomach would inactivate these enzymes if the hydrochloric acid were not neutralized by bicarbonate ions also secreted in large amounts by the pancreas. The hormone secretin is responsible for the pancreas knowing exactly how much alkalinity (bicarbonate) is needed. This alkalinity is also donated from the blood, as is the acidity for stomach acid.

The Liver and Gallbladder

Because fat is not soluble in water, the digestion of fat in the small intestine requires special processes to emulsify (degrease) these molecules. This is brought about by a group of detergent molecules, known as bile salts, that are secreted by the liver into the bile ducts, which eventually join the pancreatic duct and empty into the duodenum.

Any food high in fat must be emulsified. There are no digestive enzymes in bile. Bile is only a degreaser. Foods such as peanut butter and salad dressings are very difficult to digest because of their oil content. If the oil is not degreased, enzymes cannot penetrate the oil to digest the food. That is what bile does. So we need bile to expose the bonds within the food that the enzymes need to break. This is an important step in the digestive process. Obviously, much gas will be formed if the action of bile is not adequate.

Between meals, secreted bile is stored in a small sac underneath the liver called the gallbladder, which concentrates the bile by absorbing salts and water. During a meal, the gallbladder contracts, causing a concentrated solution of bile to be injected into the small intestine.

Also important here is the presence of sodium and water added to the bile to neutralize stomach acid. If there is not adequate acid coming from the stomach, then sodium and water are taken out of the bile and put back into the blood. For this reason, gallbladder problems and stomach acid deficiencies are often synonymous. It is important to remember that predigestion of protein in the stomach by plant enzymes creates acidity by breaking the amino acid bonds and releasing acid molecules, thus improving the flow of bile.

When water and sodium are reabsorbed from the bile into the blood, the bile is thickened and its flow is very sluggish. This allows the formation of gallstones. They grow very slowly, of course, but the symptoms of gas, bloating, pain, and constipation gradually become overwhelming.

Eventually, the gallbladder is removed to correct the symptoms of gas, indigestion, and pain. When this is done, the surgeon simply makes a direct connection from the gallbladder to the pancreatic duct. The bile continues to drip constantly from the liver, and the body actually makes another little pouch. The gallbladder is gone, but the bile keeps coming and the hormone released by the wall of the small intestine keeps stimulating for its release. Nothing changes, except the symptoms. The patient feels much better, but the bile is still too thick! More stones will form, and the symptoms usually come back in about two or three years.

The Pancreas

The hormones formed in the wall of the duodenum also signal the pancreas as to exactly how much pancreatic secretion of enzymes and bicarbonate will be needed to digest the amount of all four major food components (protein, sugars, starches, and lipids) that are leaving the stomach.

- Pancreatic protease digests the long protein chains found in meat, eggs, and cheese into smaller protein chains that can be absorbed across the gut wall into the bloodstream.
- Pancreatic amylase digests starch and glycogen, but not cellulose, to form the simple sugars, lactose (dairy), maltose (grains), and sucrose (white).
- Pancreatic lipase digests neutral fat into glycerol (to be converted into glucose) and fatty acids.

FIGURE 12

LIPID DIGESTION

Mouth

Chewing is essential to expose surface areas for enzymatic action. A weak lipase is secreted from the sublingual glands.

Stomach

Salivary lipase and any supplemental plant lipase is joined by gastric lipase to hydrolyze lipids for 30 to 60 minutes until hydrochloric acid reduces the pH of the stomach below 3.0.

Duodenum

Bile emulsifies fat, further exposing bonds that can be hydrolyzed by pancreatic lipase. Any supplemental lipase present will be reactivated.

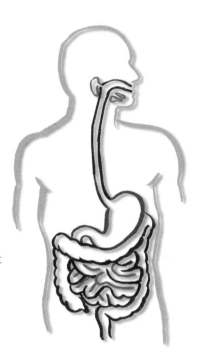

Individual fatty acids are released from their bonds with glycerol, and triglycerides, diglycerides, and individual fatty acids become available to be absorbed. Short-chain fatty acids (up to 12 carbons) are attracted to water and are absorbed directly through the intestinal wall.

Long-chain fatty acids are transported through the wall with the help of chylomicrons into the lacteals (lymph vessels). The chylomicrons are unable to enter the capillaries as do amino acids and monosaccharides. Thus fat is absorbed into the lymphatic system and then the systemic veins.

Consequently, these chylomicrons can be seen circulating in the blood following a meal high in fat. Finally, they reach the liver, the principal site of fat metabolism.

FIGURE 13

PROTEIN DIGESTION

Mouth

Chewing is essential to expose surface areas for enzymatic action. Weak proteolytic enzymes are secreted by the submandibular glands.

Stomach

Salivary enzymes and any supplemental plant enzymes hydrolyze proteins for 30 to 60 minutes until hydrochloric acid reduces the pH of the stomach to 3.0. Pepsin continues to work until the chyme is moved into the small intestine, where the pH rises above 5.0.

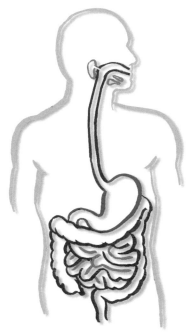

Duodenum

Pancreatic proteases continue digestive activity begun in the stomach to reduce long-chain polypeptides to short-chain polypeptides, tripeptides, and dipeptides. Many are absorbed into the blood in these stages.

Amino peptidases and dipeptidases continue to reduce peptide linkages to smaller chains and even single amino acids for absorption across the gut wall and into the portal vein.

Portal vein

Amino acids are transported to the liver, the principal site of protein metabolism. Linkages too large to be utilized by the liver must be attacked by the immune system.

The Final Step in Digestion

The second section of the small intestine, after the duodenum, is called the *jejunum*. By the time food reaches this area, the proteins and fats have been exposed to all the digestive action they are going to receive. As the food moves through this area of the small intestine, actions started by the protein- and fat-digesting enzymes from the pancreas will continue their work, and the digested molecules of food will be absorbed—all but the carbohydrates, that is. Pancreatic amylase cannot complete carbohydrate digestion. It only breaks carbohydrate molecules to the disaccharide stage or two simple sugar molecules bonded together as follows: lactose (from dairy products), maltose (from grains), and sucrose (from white sugar and flour). These sugars must be digested further in order to be available to the body as nutrients.

The *microvilli*, small fibers in the wall of the jejunum, make lactase, maltase, and sucrase to finish the digestion of carbohydrate to glucose. If these disaccharidases cannot be made in adequate amounts or if excessive amounts of these sugars are ingested, then problems result in the bowel movements of the individual.

- First, lactose and maltose cannot be absorbed unless digested. Their presence in the undigested state creates painful gas and diarrhea.
- Second, sucrose can be absorbed across the gut wall. While it cannot be turned into energy, it will cause problems in the blood by stressing the kidneys as well as producing constipation.

The final stages of digestion and most absorption occur in the small intestine. The end products of digestion—amino acids, monosaccharides, and fatty acid molecules—are now able to cross the mucosal barrier and the layer of epithelial cells that line the intestinal wall and to enter the blood and/or lymph.

FIGURE 14

CARBOHYDRATE DIGESTION

Mouth

Chewing is essential to expose surface areas for enzymatic action. Chewing releases cellulase from vegetables to remove cellulose protection and allow enzymes access to the food surfaces.

Amylase is secreted by the parotids to initiate complex carbohydrate digestion.

Stomach

Salivary enzymes and any supplemental plant enzymes continue to hydrolyze carbohydrates for 30 to 60 minutes until hydrochloric acid reduces the pH of the stomach to 3.0. Various sources claim 40% to 85% of starches can be digested before the pH reaches 3.0.

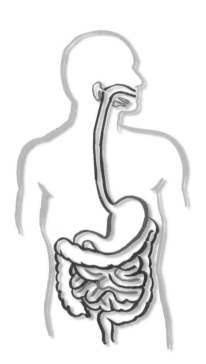

Duodenum

Pancreatin amylase completes the breakdown of carbohydrates to maltose, lactose, and sucrose. Many complex carbohydrates contain raffinose and stachyose, which the body cannot digest. This is responsible for much gas formation.

The Jejunum

It is in the middle portion of the small intestine that the final step of carbohydrate digestion occurs. The microvilli of the jejunum secrete lactase, maltase, and sucrase. The absence of any of these is capable of producing gas, bloating, and either painful diarrhea or constipation.

Absorption of Nutrients

Other organic nutrients (such as vitamins), minerals (such as sodium and potassium), and water are absorbed in the small intestine. Monosaccharides and amino acids are absorbed across the wall of the small intestine by specific active-transport processes in the epithelial membranes, as are coenzymes (vitamins and minerals).

Fatty acids enter the epithelial cells by diffusion, and water follows passively by osmotic gradients. Most digestion and absorption has been completed by the middle portion of the small intestine.

The motility of the small intestine mixes the contents of the lumen with the various secretions, bringing them into contact with the epithelial surface where absorption takes place, and slowly advances the luminal material toward the large intestine. Since most substances are absorbed in the early portion of the small intestine, only a small volume of minerals, water, and undigested material is passed on to the large intestine, which temporarily stores the undigested material (some of which is acted upon by bacteria) and concentrates it by absorbing water.

The Large Intestine

When we consider that over 90% of all cells associated with the human body are bacteria living in the large intestine, it is surprising that so little is known about what these microorganisms actually do. What we do know is that maintenance of a normal bacterial flora is imperative for health. It is now universally conceded that autointoxication (lack of adequate intestinal flora) is the underlying cause of an exceptionally large group of symptom complexes.

The colonic microflora is a dynamic population that is influenced by its host (your body), and in turn it influences you. Interactions between the microflora and the human body have implications for nutrition, infection, metabolism, toxicity, and cancer.

At the nutritional level, the bacterial population in the colon obtains all of its nutrients from the host through either *undigested dietary residues* or intestinal secretions. In return, you get back some nutrients in the form of certain vitamins and short-chain fatty acids (SCFA).

The microflora can be a source of infection found in wounds and the urogenital tract, and some bacteria can generate toxins. Nevertheless, the flora plays a very important role in preventing many pathogens from establishing in the gut and causing disease. This process is known as colonization resistance and is a major beneficial effect of the normal gut flora. This is very important in the overgrowth of fungi/yeasts, such as *Candida albicans*, in the bowel. These organisms flourish when a diet of undigested simple sugars is provided.

From the standpoint of their role in toxicity, bacteria in the colon can have both beneficial and detrimental influences. The involvement of the microflora in toxic events is often determined by the body's metabolism—in other words, by the body's ability to digest its food and eliminate the source of putrefying matter. For example, the bacteria can be responsible for the conversion of an ingested compound into a form that is more or less toxic than the parent compound and can result in activating or detoxifying the compound, respectively. There are, however, examples of more subtle effects of the flora on the toxicity of chemicals; however, the subject material available is vast and beyond the scope of this text.

Part Three

THE
CONSEQUENCES OF
POOR DIGESTION

What Happens When Food

Is Not Adequately Digested?

NSAIDs

If the building's foundation isn't level, it won't matter
how expensive the windows
are, after awhile they stick
and won't open or close.
Good health depends on good
digestion. There is no
substitute. It is not nice to
fool Mother Nature.

Chapter 7

What Happens When Food Is Not Adequately Digested?

When digestion doesn't occur normally, no attention is paid to the consequences or its restoration. Our society has come to rely on covering up symptoms of poor digestion with antacids, products to relieve the discomfort of abdominal gas, laxatives, and products to relieve diarrhea. However, little attention is paid to the prevention of these occurrences.

Dietary selections and their digestion are critical in the prevention of disease. For example, lactose-intolerant patients are advised to avoid milk and ice cream to prevent the occurrence of the unpleasant symptoms that follow their ingestion. Usually they also resort to using one of the lactase products now on the market. However, no thought is given to the fact that lactase is only one of the sugar-digesting enzymes secreted by the small intestine. The inability of that organ to make one enzyme is usually accompanied by the inability to make the others.

Having reviewed normal digestion, let us now look at what happens when food is not adequately reduced in size to be utilized for energy production by the body. Do these processes challenge homeostasis, and are they disease-productive? It is necessary to divide our studies into two groups:

- Food particles not digested well enough to pass across the gut wall pass down the alimentary canal, where they will putrefy and form toxins that will be absorbed into the blood.

- Food particles digested well enough to pass through the gut wall and into the blood but not reduced to particles small enough to be utilized by the body for energy production. Specifically, we will examine the process of digestive leukocytosis and the formation of circulating immune complexes and the resulting fibromyalgia.

Indicanuria and Bowel Toxicity

A General Indicator of the Inability to Digest Food

Inadequately digested food molecules, not digested well enough to be absorbed across the gut wall, are acted upon by "unfriendly" bacterial growth in the last section of the small intestine. You know that there are trillions of cells in the human body. But did you know that 80% of all the cells in the body are bacteria living in the gastrointestinal tract? They are in the ileum (the last section of the small intestine) and in the colon. Lactobacilli, the bacteria that retard the souring of milk, are referred to as the "friendly bacteria" that grow and prosper in the intestinal tract. You know that there are bacteria that are not friendly. What happens when the unfriendly bacteria outnumber the friendly ones? How does that happen?

The unfriendly bacteria feed on food that has not been adequately digested. Most of it is protein in nature. But the process also occurs when protein and refined carbohydrates are consumed at the same meal, and when free oils are given at mealtime to a fat-intolerant patient.

This process of indican formation is called *putrefaction*. That's a five-hundred-dollar word meaning rotten or rotting. When food rots, chemicals are formed in it that are toxic or poisonous to the body. They are so poisonous that they irritate the lining of the colon, the mucus that protects the wall of the gastrointestinal tract. These chemicals are so toxic that the body doesn't even leave them in the colon waiting for you to evacuate the bowel. By the way, the more rotting or putrefaction that's going on, the more constipated the person is.

The word *constipation* is widely misunderstood by the public. When you talk to people about being constipated, they usually have the impression that constipation means being unable to evacuate the bowel, or that it hurts, or that the stool is very hard.

Constipation should be understood to mean not having at least one bowel movement a day. Unless the constipation is quite chronic, most people deny being constipated. The best way to inquire about this matter is to ask, "How many bowel movements do you have a day?" If the reply is less than one per day, the person is constipated.

When this happens, toxic chemicals (known collectively as *indican*) are formed and they irritate the walls of the bowel. When the functioning of any aspect of the gut mucosal barrier is sufficiently compromised, the integrity of the bowel itself becomes compromised, resulting in increased permeability to foreign or gut-derived antigens, allowing them to "leak" through the gut into the lymphatics and the systemic circulation. Most of these compounds can be excreted in the feces; however, the remainder are absorbed into the blood. Since these compounds are toxic to the system, they can cause a number of inflammatory problems before they can be detoxified by the liver.

This process has been underestimated as a stress to the liver. If and when these compounds are detoxified, they are returned to the blood and eliminated by the kidneys. Acting as foreign invaders, they can target specific organs, producing pain and an inflammatory immune response enhanced by immune complex deposition. This can lead to recurrent fibromyalgia or myofascitis as well as symptoms of infection, often without an infectious agent because immune responses to foreign invaders are identical, whether infectious or not.

The Mucosal Barrier

It is obvious that reducing the permeability of the gut to foreign antigens is a primary preventive and therapeutic tool in the care of many chronic conditions. It is clinically prudent to consider autointoxication, or "leaky gut syndrome," as an integral part of any chronic condition. The ability of the body to produce and maintain normal, healthy, well-nourished mucosal cells to protect the epithelial walls of the gastrointestinal tract is imperative.

TABLE 7

SYMPTOMS OF INDICANURIA (INTESTINAL TOXEMIA)

Skin—Hair—Nails

Dermatosis

Eczema

Psoriasis

Eyes—Ears—Nose—Sinuses

Diseases of nasal accessory sinuses

Diseases of middle and internal ear

Eye strain

Genitourinary

Foul odor to urine

Cardiovascular

Tachycardia

Cardiac arrhythmias

Migraines

Endocrine System

Breast pathology

Eclampsia

Thyroid goiter

Musculoskeletal System

Arthritis

Low back pain and sciatica

Fibromyalgia and myofascitis

Respiratory System

Asthma

Gastrointestinal

Gas and bloating

Constipation

Crohn's disease

Diarrhea

Food allergies

Foul stool odor

Gastritis—Heartburn—Hiatal hernia

Inflammatory bowel disease

lleocecal valve

Malassimilation—Weight loss

Mouth—Throat

Body Odor—Halitosis

Nervous System

Depression and melancholy

Epilepsy

Excessive worry

Incoordination

Irritability

Lack of confidence

Loss of concentration and memory

Mental sluggishness and dullness

Schizophrenia

Senility

Sensory polyneuritis

The body's protective armor against invasion of bacteria and toxins through the digestive tract is composed of the following:

- Adequate hydrochloric acid/pepsin in the stomach.
- Adequate biliary secretion to emulsify the fats ingested.
- Adequate pancreatic enzyme production.
- Microvillous enzyme activity in the jejunum to digest simple sugars.
- Healthy intestinal microflora (lactobacillus).
- Healthy mucus to trap food digestive remnants and adequate mucosal, secretory antibodies to neutralize those that are immunoreactive.
- An intact barrier in the intestine to resist invasion by large molecules while allowing absorption of essential nutritive and energetic factors.
- Gut-associated lymphoid tissues of Peyer (Peyer's patches) to trap foreign invaders that manage to elude the primary trapping systems.

When the Mucosal Barrier Fails, the Immune System Must Make Up for Poor Digestion

Certain white blood cells hurry to clean up any area of inflammation caused by irritation. Other white blood cells (T-cells) defend against foreign invaders. Other specialized tissues, such as the Kupffer cells in the liver and sinusoidal cells in the spleen, remove damaged cells and immune complexes. There is also an increase in antibody production.

Digestive Leukocytosis

Leukocytosis is a pathological condition commonly found in cases of infection, intoxication, and poisoning. But did you know that in 1843 Franz Donders, a Dutch ophthalmologist, discovered that the number of blood cells circulating in the peripheral blood actually varied and was not constant? Rudolph Virchow, the father of cellular pathology, described digestive leukocytosis in 1897 and considered it to be a normal condition because all his subjects demonstrated it after ingesting food! The leukocytes are rich in enzymes and are called on to finish digestion not completed in the gut.

Obviously, they are being called on to compensate for pancreatic enzymes that could not completely digest the food that was eaten. This condition occurs after eating food that is enzyme deficient or cooked. (Cooking destroys food enzymes at 118 degrees Fahrenheit.) The body needs those enzymes to predigest food in the stomach. When they are not present, the body must mobilize its immune system to finish the digestive process. Is it any wonder that chronic degenerative diseases are not only increasing but appearing at earlier stages of life than ever before?

All cells contain digestive enzymes for every known organic compound. When foreign material comes in contact with a cell wall, the cell membrane gradually surrounds the particle and engulfs it. It then uses its enzymes to digest the material and distributes it throughout the entire cell and uses it as a food source.

While any cell can perform this function, it remains for the white blood cells to travel through the blood and destroy larger particles of matter, such as bacteria, cell fragments, or inadequately digested foods that are free in the extracellular fluid. When one of these white blood cells comes in contact with a foreign particle under appropriate circumstances, the membrane engulfs the particle and moves it to the inside of the cell, where it is digested by enzymes. Thus, the difference between this process in a white blood cell and any other cell is primarily a matter of size.

Paul Kautchakoff, M.D., expanded Virchow's findings in 1930 by proving that digestive leukocytosis was caused by eating cooked food, but not by eating raw food. Kautchakoff found that he could divide his findings into four distinct groups according to the elevation of white cells in the blood:

• Raw or frozen food produced no increase in the white blood cell count.
• Commonly cooked food caused a mild leukocytosis.

- Pressure-cooked or canned food produced a moderate white blood cell elevation.

- Man-made foods (which do not contain food enzymes), such as carbonated beverages, alcohol, white sugar, flour, and vinegar, were the most offensive, causing a severe leukocytosis.

Kautchakoff went so far as to prove that meat must be eaten raw to avoid leukocytosis and that cured, salted, canned, and cooked meats brought on a violent reaction equivalent to the leukocytosis seen in poisoning.

What segment of the population consumes the most cooked, canned meat? The usual answer is members of the military. But babies consume much more on a regular basis. Remember, this naturally protective process severely taxes the immune system and is easily correctable utilizing dietary modification and proper enzyme supplementation.

Circulating Immune Complexes (CICs)

When the intestinal mucosa becomes inflamed, it becomes more permeable. When digestive remnants enter the blood and excite the immune system, they become known as *circulating immune complexes*. Circulating immune complexes (systemic foreign antigens) are the leading cause of fibromyalgia and can retard healing, promote and prolong pain from inflammatory processes, and reduce the competency of the immune system. Only an intact intestinal mucosal barrier protects the body from entry by foreign antigens and their systemic effects. The major factors that compromise the mucosal barrier are:

- **Therapy with prostaglandin inhibitors** such as nonsteroidal anti-inflammatory drugs (NSAIDs) or aspirin and steroids such as prednisone and cortisone. Prostaglandin inhibitors suppress repair and have been shown to increase gut permeability with moderate use. Long-term steroid use can cause stomach and duodenal ulcers and immune suppression, contributing significantly to gut hyperpermeability and its complications.

- **Antacids that decrease the acidity of the stomach**, reduce the activity of pepsin, and limit the stomach's ability to adequately digest proteins. This compromise increases the number of undigested, large molecules entering the bowel and systemic circulation. By decreasing stomach acidity, antacids can also impair the absorption of minerals.
- **Antibiotics that disrupt the normal balance of bacterial microflora in the gut as well as the mouth, skin, and vagina.** This often leads to serious overgrowth of pathogenic microflora in these areas, resulting in infection and inflammation. Proliferation and overgrowth of *Candida* and other yeasts in the gastrointestinal tract can result in a complex of symptoms from gas, bloating, and gastrointestinal distress to unexplained chronic fatigue, depression, and various systemic inflammatory disorders.

Fibromyalgia

Initially, fibromyalgia was believed to be a psychiatric disease. It is now widely accepted that fibromyalgia is an independent physical illness. You will find its symptoms are strikingly similar to those of bowel toxicity. Fibromyalgia (FM) is characterized by:

- Musculoskeletal pain and aching
- Disturbed sleeping patterns
- Fatigue
- Morning stiffness and local tenderness
- Headaches
- Depression
- Paresthesia (feelings of numbness and tingling, restless legs)
- Bowel and bladder disturbances
- Subjective soft tissue swelling
- Raynaud's phenomenon

ENZYMES: *The Key to Health*

- Rhinitis (sinus problems)

- Bruxism (grinding of the teeth, especially when sleeping)

- Bursitis

- Sciatica

- Refractory allergies

- Temporomandibular joint dysfunction (TMJ)

Despite the symptoms, physical laboratory and radiologic studies are often normal because advanced immunology tests are often not performed. Unlike rheumatoid arthritis, this connective tissue disorder is not associated with deformity or inflammation of the joints. Thus, the diagnosis has been clinical rather than objective.

The obvious remedy or magic bullet for many of these symptoms is aspirin, acetaminophen, or other pain relievers. If over-the-counter drugs are not effective, then usually prescription pain relievers are used.

Chapter 8

NSAIDs

Nonsteroidal anti-inflammatory drugs (NSAIDs) are the most frequently prescribed medications worldwide. In the United States, an estimated 70 million prescriptions are written yearly. This does not include the over-the-counter market.

NSAIDs are one of the most common drug groups associated with side effects, which are well-known and predictable:

- Gastrointestinal bleeding
- Kidney failure
- Hepatotoxicity
- Central nervous system effects

TABLE 8

TOXICITY RATING OF NSAIDS ACCORDING TO OCCURRENCE OF GASTROINTESTINAL SIDE EFFECTS

	Generic Name	**Brand Name**
Least Toxic	Ibuprofen	Advil
	Diclofenac	Voltaren
	Indomethacin	Indocin
	Naproxen	Anaprox/Naprosyn
	Piroxicam	Feldene
Most Toxic	Ketoprofen	Orudis

Gastrointestinal Bleeding (Digestive Tract Damage)

The most widely known negative effect of NSAIDs is gastrointestinal irritation.

- NSAID users are four times more likely to be hospitalized with bleeding ulcers or gastrointestinal hemorrhage than nonusers.

• Available studies measure only bleeding ulcer occurrence and thus ignore patients who discontinue use of NSAIDs after experiencing gastrointestinal discomfort.

End-Stage Renal Disease (Kidney Damage)

Researchers estimate that 8% to 10% of the overall incidence of kidney failure is directly attributable to acetaminophen. The risk is dose-dependent, beginning at 105 to 365 pills per year with a sharp increase after 1,000 pills per year. Most alarming is the incidence in individuals who consumed more than 5,000 pills in a lifetime. Aspirin has no effect on end-stage renal disease. (Source: Pemeger, Whelten, and Kleg, "Risk of Kidney Failure with the Use of Acetaminophen, Aspirin, and Nonsteroidal Anti-inflammatory Drugs," *New England Journal of Medicine* 1994, vol. 331, pages 1675–1679.)

Hepatotoxicity (Liver Damage)

Studies indicate that patients using NSAIDs on a daily basis show a twofold increase in liver enzyme studies. This is most significant with the use of diclofenac (voltaren). (Sources: Robinivitz and Van Thiel, "Hepatotoxicity of Nonsteroidal Anti-inflammatory Drugs," *American Journal of Gastroenterology*, 1992, vol. 87, pages 1696–1704; Helfgott et al., "Diclofenac Associated Hepatotoxicity," *Journal of the American Medical Association* 1990, vol. 264, pages 2660–2662.

Central Nervous System Effects

The following side effects have been attributed to NSAIDs:

• Psychosis

• Cognitive dysfunction

• Hallucinations

These side effects are primarily seen in the elderly. (Source: Hoppmann et al., "Central Nervous System Side Effects of Nonsteroidal Anti-Inflammatory Drugs," *Archives of Internal Medicine*, 1991, vol. 151, pages 1309–1313.)

Inflammation and Enzymes

Inflammation is not a disease, and it should not be suppressed by the use of anti-inflammatory medications. This seemed like sage advice from *Boyd's Pathology* and my clinical instructors when I graduated from Logan College, and it has not been contradicted over the years. In fact, the side effects of NSAID use are now well understood and are a strong endorsement for the advice I received over 30 years ago. But I soon found that putting that advice into practice was difficult. All the therapeutic measures I had at my disposal were not always enough to combat the four cardinal signs of inflammation. It is true that fever, redness, swelling, and pain herald a defensive action by the body, but it is also true that they signal an increased need for nutrients during the crisis, be it infection, allergic reaction, or traumatic incident.

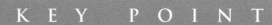

KEY POINT

Health is a community or city working together in harmony, with all groups supporting one another. Disease requires police intervention (immune system) to restore order, because the city is being overrun with criminals or invaders. If the police cannot handle the situation locally, the military (a prescription drug) is called in to assist.

The majority of doctors believe that nutritional supplementation is an important adjunct to their practice. There is arguably no greater need for additional nutrients than during an inflammatory reaction. For example, acute trauma results in a rapid loss of body protein, and the amount depends on the extent of tissue damage. Tissue protein must be replaced, and that presents problems in patients with compromised protein digestion, such as geriatric patients, for example.

During an inflammatory reaction, our bodies require increased amounts of fibrin, plasmin, thrombin, and kinin. However, the primary response to inflammation is by the immune system and its increased use of enzymes. Localized deficiencies of enzymes can prolong inflammation and delay healing. It has been reported that oral use of proteolytic enzymes can reduce healing times by up to 50%. In addition to use by the immune system, enzymes break down microthrombi and fibrin clots in the injured area. This increases blood flow, and the debris and waste products are removed more quickly. Therefore, supplementation of enzymes is warranted in these cases. All of this is fairly common knowledge among professionals. What is uncommon knowledge is which enzymes should be supplemented and in what form.

Amylases and lipases, not just proteases, also play a role in the inflammatory response. Each contributes specific functions to assist the inflammatory response. These enzymes can be gathered from several sources and concentrated for supplemental use. Fungal enzymes from various species of Aspergillus work in broader pH ranges than those taken from beef and pork (pancreatin) and therefore find more frequent clinical use. This is an important consideration since there are specific pH changes in tissue during inflammation, the affected tissue becoming either more alkaline or more acid than normal. But perhaps even more important is that plant enzymes have a broader range of substrates (specific substances) they can be employed against. Digestive enzymes from the pancreas are very selective in the bonds they can hydrolyze (digest).

The use of plant enzymes is sometimes confusing. The term means the enzymes were grown on and separated from plants, not animals. But not all plant enzymes have identical properties. For example, bromelain from pineapples and papain from papaya are frequently used by supplement manufacturers. This is because they are commonly used for purposes other than dietary supplements and are therefore relatively inexpensive. Papain is used to tan leather, and like bromelain, it is used as a meat tenderizer in restaurants. Both of these

ENZYMES: *The Key to Health*

enzymes enjoy their peak rate of activity in temperatures above that of body temperature. Bromelain works best in a temperature range of 120 to 160 degrees Fahrenheit! That is why it can be used as a meat tenderizer. Papain has a similar action in temperature ranges above those of the body. Both of these enzymes will work in the human body, but not anywhere near their peak capacity. The question is, why are they used in supplements? Plant enzymes have their peak activity range between 95 and 105 degrees Fahrenheit, well within body temperature range.

This means that using enteric coated enzyme tablets between meals to protect the enzymes from hydrochloric acid is a waste of time and money. Besides, enzymes lose from 40% to 60% of their potency by being compressed into tablet form! There are a lot of other misconceptions about enzyme use.

Let us now turn our attention to normal immune functions of the body, and how enzymes play such an incredibly important role in protecting the body and why their functions should not be suppressed by the use of pain relievers such as aspirin and NSAIDs.

White Blood Cells

White blood cells are usually divided into two groups: those that contain granules that can be stained for identification and those that do not contain granules. It was established in the 1960s that the granules are actually organelles (lysosomes) containing hydrolytic enzymes.

Granulocytes
- Neutrophils are phagocytic, functioning in the destruction of pathogenic microorganisms and other foreign matter.
- Eosinophils apparently phagocytize antigen-antibody complexes.
- The function of basophils is still uncertain.

Nongranular Forms

- Lymphocytes: A somewhat heterogeneous group of cells important in the process of immunity-producing antibodies and other agents involved in the immune process.

- Monocytes: They function as phagocytes, transforming into macrophages after invading infected sites.

Macrophages

Once they have left the circulatory system and entered the tissues, the lymphocytes and monocytes progressively swell during the next few hours and become macrophages and become more powerful phagocytes than the neutrophils. They have the ability to engulf much larger particles and often five or more times as many particles as the neutrophils. And they can even phagocytize whole red blood cells or malarial parasites, whereas neutrophils are not capable of phagocytizing particles much larger than bacteria. Also, macrophages have much greater ability than neutrophils to phagocytize necrotic tissue, which is a very important function performed by these cells in chronic infection.

KEY POINT

The first rule of waging war (inflammation) is to establish a line of supply for the troops. For the body, the line of supplies would include an increased amount of nutrients to meet the increased demands.

Phagocytosis

When these cells come in contact with extraneous particulate matter either in the tissues or in the bloodstream, the phenomenon of phagocytosis takes place extremely rapidly, with the particle passing through the cell membrane to the inside of the phagocyte within a few hundredths of a second.

Obviously, the phagocytes must be selective; otherwise some of the structures of the body itself would be ingested. Whether or not phagocytosis will occur depends especially on three selective procedures:

- First, the body has a means of selectively combining foreign particles with globulin molecules called *opsonins*. After the opsonins have combined with the particle, the globulin allows adhesion of the phagocyte to the surface of the particle, which promotes phagocytosis.

- Second, if the surface of a particle is rough, the likelihood of phagocytosis is increased, whereas a smooth particle is very resistant to phagocytosis.

- Third, most natural substances of the body have electronegative surface charges and therefore are repelled from the phagocytes, which also carry electronegative surface charges. On the other hand, dead tissues and foreign particles are frequently electropositive and are therefore subject to phagocytosis.

Enzymatic Digestion in the Phagocytes

Once a foreign particle has been phagocytized, the cell immediately begins digesting the particle. Neutrophils and macrophages both have proteolytic enzymes especially geared to digesting bacteria and other foreign protein matter. The macrophages also contain large amounts of lipases, which digest the thick lipid membranes possessed by tubercle bacteria, leprosy bacteria, and others. Macrophages are capable of ingesting four times as much material as neutrophils before dying.

Pinocytosis

The cell membrane has the ability to imbibe small amounts of substances from the extracellular fluid by the process called *pinocytosis*. This occurs particularly when large quantities of proteins or excessive amounts of salts are present in the surrounding fluid. Protein, for instance, becomes adsorbed to the membrane, which causes the membrane to invaginate and to pinch off inside the cell to form a pinocytic vesicle. Then the pinocytic vesicle combines with one or more lysosomes that discharge hydrolytic enzymes into the vesicle, which in turn digest the protein and other substances in the vesicle. The end products of digestion then become distributed throughout the cell. The real importance of pinocytosis to the body is that this is the only known means by which very large molecules, such as those of protein, can be transported to the interior of cells.

Lysosomes

The name of this organelle (Greek *lysis*, a loosening, Greek *soma*, body) suggests its nature: a body containing digestive enzymes. The enzymes present in these structures are described as hydrolytic (Greek *hydro*, water) since they break down organic compounds by the addition of water. More than a dozen such enzymes acting on virtually all classes of organic substances have been identified in lysosomes.

Lysosomes are sacs in the cell membranes that contain hydrolytic enzymes active at an acid pH that serve to digest exogenous material.

Lysosomes are responsible for a number of cellular functions, including:

- *Digestion of substances taken into the cell following phagocytosis:* A process in which the cell membrane forms a pocket enclosing the particle, which pinches free and then fuses with a lysosome. The digested particle may be a source of nutrition but more commonly is a potentially harmful agent. Lysosomes are especially prominent in certain types of white blood cells and in macrophages, large phagocytic cells located outside the bloodstream.
- *Autolysis:* The self-destruction of cells after rupture of lysosomes. An example of this is the destruction of a structure formed during each menstrual cycle, the *corpus luteum*, that secretes female sex hormones.
- *Autophagia:* The digestion of organelles or parts of organelles taken into the lysosomes during starvation.

ENZYMES: *The Key to Health*

• *Destruction of extracellular matter by discharging lysosomal enzymes to the outside of cells:* An example is the release by bone cells of enzymes that break down limited areas of bone matrix. This releases calcium from the bone and is one mechanism for maintaining normal levels of blood calcium. Release of lysosomal enzymes to the exterior of cells sometimes has harmful consequences. In arthritis, for example, the inflammation and tissue damage are caused by lysosomal enzymes. These enzymes are released to remove the circulating immune complex materials that have precipitated from the blood into the tissues. Aspirin and cortisone, used to treat arthritis, are believed to retard lysosomal enzyme activity, as discussed on page 65.

You can readily see the importance enzymes play in protecting the body from foreign invaders such as bacteria and parasites, through their functions not only in the white blood cells, but in every cell of the body.

However, no discussion of immune function would be complete without a mention of how the reticuloendothelial system protects the body against inadequately digested food.

The Reticuloendothelial System

Most textbooks list the 10 major organ systems of the body that maintain homeostasis, but few mention specific examination procedures for the reticuloendothelial system. That is because this very important system is composed not of visceral organs but rather macrophages (white blood cells) that are found in connective tissue. They carry on the process of phagocytosis or detoxification in critical areas of the body.

You will recall that 60% of body weight is water, and of that, two-thirds of this fluid is found inside the cells. The remaining one-third of total body water is found outside the cells and is referred to as the extracellular fluid (ECF). It is only the ECF that is involved with maintenance of homeostasis, that critical process controlled by the autonomic nervous system and endocrine system through the direction of the hypothalamus. The ECF is composed of the fluid in the bloodstream (20%) and the fluid outside the bloodstream (80%) in the connective tissues. The macrophages in the fluid of connective tissues comprise the reticuloendothelial system, or the macrophage system, as it is sometimes called.

Obviously, since we are dealing with cells *outside* the bloodstream, we cannot examine the blood to get definitive answers. Nor will urinalysis yield much useful information. X-rays, MRIs, and CAT scans are useless.

Macrophages are defined as mononuclear phagocytes found in the tissues. These cells arise from stem cells in the bone marrow. They develop into monocytes and circulate in the blood for about 40 hours. Monocytes are nongranular white blood cells that function as phagocytes in the blood. They are transformed into macrophages after leaving the bloodstream and moving into the connective tissues. Here they undergo a metamorphosis, increasing in size, phagocytic activity, and *lysosomal enzyme content.*

The lysosomal sacs of these cells contain hydrolytic (digestive) enzymes that are active in an acid pH. This is an incredibly important point if you are using enzymes to enhance immune function. Pancreatic enzymes are not active in an acid pH environment, only in the alkaline environments found in the blood and small intestine. The white blood cells are the mainstay of your immune system, and in particular, the reticuloendothelial system:

- Macrophages are most frequently concentrated in the loose connective tissues, which are especially numerous in mucous membranes of the digestive and respiratory tracts and in association with the lymphatics in the connective tissue of the pleura and peritoneum.
- Macrophages lining the blood sinuses of the liver (Kupffer cells) and bone marrow.
- Macrophages lining the sinuses of lymph nodes and the spleen.
- Monocytes in the bloodstream.
- Aggregated lymphatic follicles found in association with small blood vessels

and the lymphatics of subserous connective tissue of the peritoneum.

- Microglia of the central nervous system. Microglia are small, non-neural, interstitial cells of mesodermal origin that form the supporting structure of the central nervous system.

Once in the connective tissues, the macrophages become powerful scavengers, even more powerful than neutrophils. They have the ability to engulf not only much larger particles but also larger quantities than the neutrophils. As mentioned before, macrophages are capable of ingesting four times as much material as neutrophils before dying. They can even phagocytize whole red blood cells and malarial parasites, whereas neutrophils are not capable of phagocytizing particles much larger than bacteria. Also, macrophages have a much greater ability to phagocytize necrotic tissue.

It is often said that the best defense is a good offense. Certainly this is true of our own immune systems. It is easy to see how extensive is the body's use of aggregated lymphatic follicles and their complement of white cells that use hydrolytic enzymes as their chief method of attack.

I began this book by speaking of Dr. Edward Howell and his many contributions to the theory of food enzymes. Figure 15 depicts their distribution and how incredibly important they are to the maintenance of your health and well-being. It is amazing that these essential and vital nutrients should be removed from our food supply without provisions made for their supplementation.

FIGURE 15

WHERE HYDROLYTIC ENZYMES ARE FOUND

PLANTS
Ripening Process—Digestion

Predigestion in Humans

DIGESTIVE TRACTS
Salivary Glands

Stomach

Pancreas

Small Intestine

WHITE BLOOD CELLS
Digestive Leukocytosis—Phagocytosis

Inflammation

Lymph system cleans up what not digested

EVERY CELL WALL
Lysosomes

INSIDE EVERY CELL
Catalase

ENZYMES: *The Key to Health*

FIGURE 16

PRINCIPLES OF ENZYME REPLACEMENT NUTRITION

NUTRITION is the science of food: how it is ingested, digested,
absorbed, transported, utilized, and eliminated.

CLINICAL NUTRITION must be founded on the use of food
to prevent chronic degenerative disease.

Food requires digestion.

The ability to digest food is completely dependent on the
ECF to supply its needed acidity and alkalinity.

Plant enzymes are the only supplement available that will
digest food past an incompetent digestive system and can therefore
be used to normalize homeostasis and thereby favorably influencing:

DIGESTION
ACID-BASE BALANCE
ELECTROLYTE BALANCE
FLUID TRANSPORT MECHANISMS
CONNECTIVE TISSUE
and MUSCLE CONTRACTION REFLEXES.

To find the proper plant enzyme formula to use,
look at what you eat and especially at what you crave.

FIGURE 17

FIND THE CAUSE AND THE SOLUTION IS EVIDENT

Case History	=	Signs and Symptoms Survey
Dietary Analysis	=	Menu Selections
Physical Exam	=	Palpatory Diagnostic Reflex
Laboratory Workup	=	24-Hour Urinalysis
Recommendations	=	Enzyme Replacement Nutrition

THE FUTURE

*The Changing Face of
Health Care in America*

*What Is Nutritional and
What Is Pharmaceutical?*

A DIFFERENT PERSPECTIVE

In the 21st century, the science of nutrition will rise majestically in the healing arts like a breathtaking skyscraper, designed by those with the vision to use food as their building materials and employ enzymes to do the work.

Chapter 9

The Changing Face of Health Care in America

In accordance with the table of essential definitions on page 19, the secret of a successful nutritional program lies in using whole foods and even whole herbs as sources of nutrients. To ensure proper assimilation and utilization of the nutrients, our bodies need the enzymes contained in those foods to assist us in receiving maximum nourishment in the event that we do not have perfect digestion.

The process of nourishing human beings has worked wonderfully well for thousands of years. But today the process is threatened by our increasing need for longer shelf life. This need makes us dependent on processed foods. By that I mean foods that do not contain the indigenous enzymes they were created with. For whatever reason, the enzymes have been removed by cooking, canning, or irradiation, or perhaps the food was man-made and never contained enzymes.

So the question becomes this: "How do we as a modern society solve this perplexing dilemma?" The Surgeon General has already established that the increasing incidence of chronic degenerative diseases such as heart disease, diabetes, and cancer can clearly be prevented by eliminating dietary indiscretions. On the one hand, we must remove enzymes from food, and on the other, they appear to be essential nutrients.

You will not find much information on this concept since those in both traditional medicine and so-called alternative medicine are much more interested in studying active ingredients found in the food or herb. This is the magic bullet approach to preventing and healing disease that we mentioned earlier. We might very well ask ourselves, "What exactly are these active ingredients?" Because it appears they are the future, of not only medicine but nutrition as well.

The term *active ingredient* means a single chemical compound that can be isolated out of a food or herb and then concentrated into a larger dose than could be gained from the food itself. This could then be used to relieve a symptom or treat a disease. The symptom or disease, as we know, is a deviation from normal health (see Table 1 for a definition of health) and therefore evidence that the body is unable to maintain normal function. It indicates only that the sufferer is moving through the demilitarized zone of exhaustion and not necessarily that the patient is diseased. It remains to be proven that the patient is indeed diseased and nutrition would not restore normalcy.

We spent a great deal of time in the beginning of this book differentiating nutrition from medicine and food from drug. We must now ask ourselves the following question: "Is the use of concentrated chemical compounds the practice of medicine or the practice of nutrition?" The answer may be clear when speaking of the latest blood pressure medicine, anticancer drug, or even aspirin. But the issue, for many, seems to get clouded when speaking of purported natural things like vitamins and minerals. Recall that by definition, any concentrated chemical compound is a drug and not a food.

Clearly this world needs active ingredients (drugs). Equally clear is the fact that we need people extensively trained in the use of these concentrated active ingredients (prescription drugs) to ensure the protection of the consumer. Hippocrates's admonition, "first do no harm," is as relevant today as when he said it. This is made all the more clear when we learn of the danger associated with drugs.

The following information was taken from an article entitled "Doctors Gamble with Rx Drugs" published in the June 1998 issue of *Business & Health*:

> Nearly six in 10 of the 250 physicians who completed self-
> assessments for the American Medical Association admitted
> they rely solely on claims made by manufacturers' sales reps
> when prescribing new drugs or using new medical devices.

And nearly one in seven said they don't refer to the
Physicians' Desk Reference (PDR) or any other resource
when they write prescriptions for other unfamiliar drugs.

The article went on to say that a team of University of Toronto researchers reported that adverse drug reactions, in hospitalized patients, was the fourth leading cause of death in the U.S., after heart disease, cancer, and stroke. That conclusion was reached by counting only medications that were properly prescribed and administered. Drug reactions associated with drug abuse, noncompliance, overdose, or therapeutic failure were excluded.

This fact has been responsible, at least in part, for the enormous and growing interest by the public in alternative and complementary medicine.

Medicine and pharmaceutical drugs have dominated the healing scene in the 20th century. However, the dominance has been slipping very gradually for the last half of this century, imperceptibly at first. Nutrition gained a foothold in the late 1940s and 1950s. Many Americans were branded as "health nuts" because they actually felt that taking vitamins and minerals as dietary supplements was worthwhile, if only as an insurance measure. They would defend themselves by explaining, "After all, it can't hurt." Gradually, over the years, they have become bolder, even vociferous.

The alternative health movement got a big boost in the 1970s when President Nixon went to China. Americans were amazed when James Reston, a reporter on the trip, was stricken with acute appendicitis, and acupuncture was used as the only anesthetic during the surgery. There are now more than 40 acupuncture colleges in the United States.

Slowly but surely, Americans began to discover that healing didn't have to come out of a bottle or an operating room. More and more, they became aware of the side effects and dependency drugs induce, and consumers began looking for less risky alternatives, such as chiropractic, nutrition counseling, and homeopathic remedies as well as acupuncture and other therapies.

Consumers were demanding services medical doctors could not provide. By 1994 it was estimated that 30% to 40% of consumers had visited alternative practitioners. Medicine became so alarmed that, in 1994, the *New England Journal of Medicine* commissioned a telephone interview study of 1,500 persons who had sought such care (see Table 9). The study found that people were willing to pay for alternative care out of their own pockets rather than accept ineffective medical care that was covered by a third party (i.e., HMOs).

The complaint was that medical doctors were very "crisis oriented" and not trained in natural therapies that were effective for the more chronic, less urgent problems that many people must deal with. The record seems to indicate that medicine's track record in life-threatening emergencies, surgical procedures, and technological advances is laudable. But its record in chronic degenerating diseases is abysmal.

TABLE 9

TEN REASONS PATIENTS SEEK ALTERNATIVE HEALTH CARE

New England Journal of Medicine Survey

Reprinted in *Oriental Medicine*, Vol. 3, No. 2, Summer 1994

1. Back Problems	6. Depression
2. Anxiety	7. Arthritis
3. Headaches	8. High Blood Pressure
4. Sprains or Strains	9. Digestive Problems
5. Insomnia	10. Allergies

Health Food Business reported in its December 1996 issue that:

> These are the problems that doctors don't have the time, nor the inclination to deal with … but health food store personnel do. People are turning to natural remedies, vitamins, herbs, and diet for a healthier way of life. Many of these people end up in the local natural foods store by default. They simply have nowhere else to turn.

> Before the advent of the chain drug store, people would go to their neighborhood pharmacist, who knew them by name. They turned to the pharmacist for advice on which medication to take, if they could split the dosage, what side effects to expect, and other important details.

> In many communities the health food store has partially replaced the pharmacist in that role. We're also seeing many drug stores adopting the health food store concept, whereby they are selling natural remedies side-by-side with OTCs [over-the-counter remedies].

With increasing loss of revenue, medicine is faced today with a fight-'em-or-join-'em situation. In September 1998 medicine struck back with an editorial in the *New England Journal of Medicine* entitled "Alternative Medicine No Longer Should Get 'Free Ride.'" The editorial stated in part that "the increased interest in alternative and complementary medicines has resulted in a reversion to irrational approaches to medical practice, even while scientific medicine is making some of its most dramatic advances."

Attacking alternative medicines, the editorial comments: "It is time for the scientific community to stop giving alternative medicine a free ride. There cannot be two kinds of medicine—conventional and alternative. There is only medicine that has been adequately tested and medicine that has not, medicine that works and medicine that may or may not work.

"Once a treatment has been tested rigorously, it no longer matters whether it was considered alternative at the outset. If it is found to be reasonably safe and effective, it will be accepted," they suggest.

The authors concluded: "Alternative treatments should be subjected to scientific testing no less rigorous than that required for conventional treatments."

The editorial accompanied the issue's lead study, which showed a popular herbal tincture had a strong estrogenic activity against prostate cancer in eight men, but also caused significant side effects. The issue additionally contained five other reports on adverse effects associated with the use of alternative medicines.

There can be little argument against the position taken by the editorial. Both prescription medicines and alternative medicines have serious side effects and appear to be drugs. However, we have to ask ourselves whether the article is addressing the use of dietary supplements (food) or pharmaceutical nutrients (drugs). I think that an unbiased examination of the vitamin and mineral supplements on the shelves are in reality isolated chemical compounds, or drugs fully capable of harmful side effects. Eli Lilly, the founder of one of the nation's best-known pharmaceutical manufacturers, once said, "A drug with no side effects is no drug at all." I might add that there's no difference between a drug and a poison except for the dosage given.

Pharmaceutical Nutrition?

Where do these arguments leave us with regard to maintaining health versus treating disease? It is true that only medical doctors can treat disease, and whether a magic bullet for relief of symptoms came from an alternative source is irrelevant, as the *New England Journal of Medicine* editorial stated. Nevertheless, it should be pointed out that historically many medical advances have come from outside the medical profession, and it appears that the so-called alternative movement has caught organized medicine napping. Witness the many advertisements for prescription drugs that now appear on television.

Just a few years ago, it was totally unnecessary for a pharmaceutical company to have to spend money to tell the public to ask their doctors for a prescription drug.

One thing is certain: a global trend is in the making that is propelling millions of people away from traditional allopathic solutions and toward wellness and holism. This revolution is being fueled by the aging baby boomers, who dread the possibility of languishing years away as inmates in nursing homes. This segment of the population will spend increasing amounts of money trying to hold onto their youthful vigor while maintaining and improving their health. Increasing air and water pollution have also made boomers and other population segments more health-conscious.

Fortune 500 companies know that the future belongs to those who can identify emerging trends and seize them. Let's take a look at one of the products large companies have recently offered to the public under the banner of alternative or natural remedies. It is interesting that it is not nutritional, but rather an isolated and concentrated chemical compound that may have a distant cousin someplace that once was related to food. It is not by definition nutritional in nature. That would seem to indicate that once medicine accepts it, they may be placed on a prescription basis to protect the public.

The following information was taken from "AKPharma Introducing Prelief Food Acid Neutralizer," FDC *The Tan Sheet*, Vol. 4, No. 22, May '96, p. 14. *The Tan Sheet* is a biweekly publication devoted to reporting government actions regarding the health food and supplement industry.

Food Acid Neutralizer

A product referred to as a "food acid neutralizer" is being introduced as a *dietary supplement* that can be swallowed in tablets or dissolved in food prior to eating. The product contains 65 milligrams of calcium glycerophosphate. The company appears to be positioning the product as an alternative to antacids and the new stomach acid blockers by stressing that it "does not focus on treating the acid in the individual, but instead works by treating the acid in foods."

It is not my intention to be sarcastic here; I wish only to point out that a dietary supplement by definition should add to or complete dietary intake, not chemically alter the food. A more serious question is, whether this new product is nutritional or pharmaceutical in nature? This takes us back to Table 2 and Hans Selye's explanation of the general adaptation syndrome. Nutrition is well defined and its ground well established legally by the federal government. Medicine is well defined and regulated in all 50 states. The question is, who controls the demilitarized zone, or no-man's land? Who is entitled the reap the financial rewards for relieving symptoms for which no measurable disease entity can be attached?

TABLE 10

THE DIFFERENCE BETWEEN NUTRITION AND MEDICINE

NUTRITION	*NO MAN'S LAND*	*MEDICINE*
uses FOOD	Both FOOD and over-the-counter DRUGS are used	uses prescription DRUGS
to maintain HEALTH	to relieve SYMPTOMS	to treat DISEASE
represented by	which evidence	represented by
NORMAL PHYSIOLOGY and function	STRESS on normal functions	ABNORMAL PHYSIOLOGY (Pathology)

The answer to this question may seem rhetorical, but I can assure you that it will be decided in the reasonable near future. The answer will determine who controls health care in this country for the 21st century. It is one of the most important political battles that will be waged in the early part of the next century. It probably will rank just behind the economy, world population, world pollution, and world hunger. Unfortunately, at the present time, it appears the people of the United States are impervious to the situation.

Chapter 10

What Is Nutritional and
What Is Pharmaceutical?

The science of pharmacology depends on finding isolated chemical compounds (the active ingredient) to use in the treatment of illness. In the science of nutrition, protein, carbohydrates, fats, fibers, vitamins, and minerals are the active ingredients used to maintain health. All that is required for their use is adequate digestion.

We have already discussed that the appearance of symptoms signifies that the body is unable to maintain homeostasis (see Chapter 2). In other words, some organ or tissue is stressed and unable to perform its responsibilities in that regard. Since specific components of the digestive system come from the blood (see Chapter 5), it is probable that in any condition of stress, digestion is compromised.

Because this is true, supplement companies spend large amounts of money finding new technologies to move isolated chemical compounds (called food supplements) across the gut wall and into the blood without relying on the process of digestion. For example, we find terms such as *chelated* and *colloidal* being used to describe the increased effectiveness of some products.

It is interesting to learn that chelation means to attach some nutrient to a mineral in an effort to carry that nutrient into the blood. The term colloidal refers to an aggregation of molecules in a finely divided state, dispersed in a gas, liquid, or even solid material that resists settling or being filtered out. My point is that food is already chelated and in a colloidal state. It is a perfect carrier for nutrients. All that is required for their utilization is adequate digestion.

This brings us to a very important point concerning the use of isolated chemical compounds taken from food.

It is a scientific certainty that the *part* does not have the same chemistry, effectiveness, or reactivity as the *whole*. Taking one or two chemical entities from a plant and discarding the remainder as having no therapeutic value denies the basic tenets of chemistry.

Any time you extract the active ingredient from food and supplement it as an individual stand-alone unit, you create deficiencies of the synergistic elements that are in the food that the body needs to metabolize the active ingredient. Witness the fact that refined sugar use creates deficiencies of vitamin B and potassium, among other things.

In the science of nutrition, protein, carbohydrates, fats, fibers, vitamins, and minerals are active ingredients. And they are already chelated and in a colloidal suspension just waiting to be released by the process of hydrolysis (digestion).

The reason technological advances in supplement delivery systems have attracted attention is that digestion cannot be counted on to deliver the nutrients into the body.

The Food and Drug Administration and organized medicine often issue warnings about herbs and how dangerous they can be for public consumption. This position requires some explanation. Pharmacology is concerned only with the effect of the active ingredient. Since these active ingredients are no longer combined with protein, carbohydrates, lipids, synergistic enzymes, and coenzymes in the food, they are by definition no longer food and are quite capable of producing side effects so often associated with prescription drug use.

Finding a chemical in a plant and testing a concentrated dose of that chemical only, and then declaring that the plant is toxic without evaluating what the other chemical entities present in the plant may do to change, minimize, enhance, or even block altogether the action of the chemical, is not nutritional science. Refer back to scientific method in Chapter 1.

A compromise between the pharmaceutical quest for drugs and nutritional quest for foods is needed. We must blend theory not only with physiology and biochemistry, but with common sense. Theoretically, the *fresh* plant, straight from the garden or field, would seem to have the most effective and pure applicability to human health. Such a plant would be unspoiled by any processing and intact in its contents. Many medicinal plants are best in the *fresh state*, especially the culinary herbs. The aromatic oils and oleoresins are actively present, and any water-soluble vitamin loss is minimized.

But other plants and herbs are too strong to be used in the fresh state, but in the *dried state* are quite safe for human consumption. It is a basic principle of botany that the drying of plants reduces alkaloid activity, reducing their content further the longer the drying proceeds. Thus, the use of drying renders some plants, which would otherwise be far too active to use at all, tame and safe.

Health-conscious Americans are now reasonably familiar with terms commonly used to describe such things as proteins, carbohydrates, saturated and unsaturated fatty acids, fiber, vitamins, minerals, and even to some extent enzymes. But they are not as well versed in terms used in herbology to describe other compounds found in foods and herbs. So it behooves us to take a short course in herbology. Let's start with the most frequent term used by pharmacologists to declare herbs unsafe.

Alkaloids

The term *alkaloids* describes a class of substances that are nitrogenous (containing nitrogen, as all protein-containing plants do) and have the potential to produce alkaline reactions in various body processes. Since the blood must be maintained in a very limited pH range, any substance that can push the pH in a more alkaline direction can be potentially dangerous.

Alkaloids are found to a greater or lesser degree in most medicinal plants, and traditional pharmacy in the latter part of the last century decided to call them the active ingredients of each plant and began to chemically isolate and extract them.

You will readily recognize the more common alkaloids. For example, there's morphine, heroin, and codeine, all alkaloids from the opium poppy. There's hyoscyamine, hyoscine, and scopoloamine, extracted from helleborc, henbane, and so on. There's reserpine from snakeroot (*Rauwolfia serpentine*), and atropine from belladonna, the deadly nightshade, as well as ephedrine from ephedra (*Ephedra chinensis*) and digitalis from foxglove.

Once these alkaloids were chemically identified, it became possible to synthesize them in the laboratory and manufacture them far more cheaply than the cost of extracting them from plant sources. Modern pharmacy had begun, and phytopharmacy (the chemistry of whole, real plant parts) began to decline.

A whole range of potentially deadly substances had been separated from nature's more balanced chemical partnerships and could be injected into and ingested by humans in the name of health improvement! Other chemical identities in the plants, like proteins, carbohydrates, lipids, vitamins, and minerals, were discarded altogether as not being involved in any beneficial effectiveness at all—an astonishing conclusion! Worse, the effects of these alkaloids, unmitigated by the synergistic chemistry of the original plant, could be proved by example and used to react more dangerously than the original whole plant! The part is not the same as the whole, and never will be, in any of the true sciences.

Extracting these alkaloids or making them synthetically breaks one of the first rules of nature—*nothing in nature works alone*. But in the field of pharmacology, this is necessary, as was explained in our discussion on the scientific method in Chapter 1. While extracting these alkaloids and concentrating them for medicinal use may have proved to be good pharmacology, the information gained has nothing to do with nutrition and should not be used to regulate foods and food consumption.

ENZYMES: *The Key to Health*

Glycosides

Glycosides (sometimes spelled glucosides) yield various sugars on hydrolysis (digestion). Because of this, they are both energy-producers and laxatives. The tonic and restorative effects of glycosides are responsible for many good reports from patients after using products containing them.

A lift of energy, physically and mentally, is often noticed, and they provide a stimulant effect, especially in muscles. The name glycoside also refers to the muscle sugar *glycogen*, which is made by the liver and distributed out to muscles for energy. It is necessary to have good dietary intakes of potassium in order to begin the processes of chemical change required to break down glycosides into their final state of use. Fortunately, nature has already put the two together, and these plants are already high in potassium and glycosides. Interestingly, pure white sugar has such impurities as potassium removed, and overconsumption of white sugar produces mild symptoms of potassium deficiency, such as constipation, as is well known by parents, especially around such holidays as Christmas, Easter, and Halloween.

Sugars in general provide not only energy but also waste removal stimulus. More energy to the intestinal tract and bowels improves peristalsis, the contracting rhythmical movements that propel food onward and wastes outward. Glycoside-rich plants will also be useful for the liver as laxatives and digestive stimulants. Since the heart is also a muscle, glycoside-rich plants maintain strong heart function. Medically, cardiac glycosides from plants like *Digitalis lanata* produce the drug lanoxin, a standby for many heart patients.

There are many glycosides whose names you will recognize. *Vanillin* is one that may surprise you. It's not just a delicious flavor, but a tonic and a stimulant to muscles and the digestive system as well. Of course, this applies only to the *real* vanilla pod soaked in a little milk, or soy milk, for an hour or two before use. The synthetic vanilla essence may fool you on taste, but your liver won't be fooled by its chemistry.

Amygdalin

Amygdalin, from almonds and other stone-fruit kernels like the apricot, is also a glycoside. Its use, either synthetically copied or isolated from apricots, was in brief favor recently as yet another questionable breakthrough for cancer prevention. Like any other glycoside, amygdalin can stimulate waste removal and improve general energy, and can therefore be useful for removal of *any* disease process from the body. Any means at all of raising energy and improving elimination will improve the body's ability to fight the processes that are the true nature of disease.

Saponins

The Latin word *sapon* means "soap." Saponins are a class of glycosides. As you may know, the hydrolysis (digestion) of fats is accomplished by lipase and yields fatty acids plus glycerol. The hydrolysis of a fat in the presence of an alkali is called *saponification*. This process yields glycerol and the alkali salt of the fatty acid (soap).

• Glycerol has definite nutritive value. It has the same caloric value as sugar and probably follows a similar course when utilized by the cells.

• Soaps are cleansing agents because they emulsify (degrease—break into small globlets), and the emulsification of fats in the intestine is essential for the digestion of food.

Saponins are therefore important to us as a source of energy and as a source of fatty acids. In addition, they improve biliary function (emulsification or degreasing of fats) and have a mild laxative effect.

Tannins

There are tannins to a greater or lesser degree in many plants. Tannins are used industrially just as the body tissues use them—as an astringent.

Astringents dry, tone, and tighten. All tannins are tonic in effect. Fluids are either retained better in tissues or forced from areas of sluggish fluid retention and allowed to drain better through elimination channels.

Tea contains tannins. In accordance with nature's unbending major law, one or two cups of tea act as a tonic, a gentle diuretic, and a diaphoretic. Too many cups and this process is reversed. Eighteen cups of tea a day, and you're in big trouble—not just from cancer, but from any disease at all because of unbalanced consumption of any one kind of food or fluid.

Almost any given substance can produce the irritation, inflammation, and immune response that can begin a cancer process. Tannins don't do it any more than peanut butter or hydrocarbons: too much tannin, too much peanut butter, too many petrol fumes, cigarettes, or asbestos fibers, or too much of anything can cause imbalance and produce conditions where the random cells that everyone is born with clump together and begin a cancerous process. The immune system's attention is elsewhere fighting off these repeated assaults of a local inflammatory kind.

Most tannin-containing plants are sharp-tasting and refreshing. Like a dry wine, the effect is to contract tissues. If you become violently thirsty after a dry Bordeaux red or wake at 4 a.m. and drink water copiously to lubricate your dry mouth and contracted tongue, it's the tannins in the wine that produced this.

Tannins are complex compounds, capable of combining well with proteins. Sometimes they combine too well, making the protein somewhat indigestible; at other times, and in different quantities, they enhance protein digestion. Do you like to drink fluids with a meal? Or do you prefer a drink half an hour before or after? Herbs containing tannins may be the corrective treatment if you suffer either way from too much, or too little, salivary and digestive fluids, especially if this is most obvious when eating or drinking.

Herbs that stop external fluid loss from bleeding, or plasma loss in burns, even severe hemorrhages internally or externally, are always high in tannins.

So there really was some sense in that old folk remedy of cold tea applied to burns to dry the tissue and stop fluid loss.

There are many other types of plant contents, and our short list could become very long. There are as many starches and fats, oils and resins, and saponins as there are excellent textbooks available, full of long, difficult names and biochemical chains of activity and interaction. I will recommend these to readers who can deal with the technical aspects. These books are listed in the Herbal Bibliography on page 165. For those readers who look for an overview and simple explanation, let's press on.

Steroids

The term *steroids* may make us think more of athletes and training regimes than herbal medicine, or it may bring to mind the various pharmaceutical preparations of cortisone available. Steroids are hormonal substances, and they are also found in the plant world. Phytosteroids resemble our human hormones like twins resemble each other: they may do different things at different times, but they are virtually identical in characteristics and structure.

As is usual with pharmacy, extraction of one steroid or another can be made and then prescribed on its own. I prefer nature's balanced ingredients used together. Many medicinal herbs contain one or more of the steroids estrogen, progesterone, testosterone, cortisol, and vegetable forms of these and other hormones that act like the human forms, although they are not identical to them. Lanolin is an animal hormone that is easily absorbed by human skin. It is used as a base for mixing herbal extracts into ointments.

Perhaps our most notorious steroid today is cholesterol. Every second human, male or female, seems to want to reduce cholesterol. You should know that cholesterol is made by the liver, is used in seminal fluid and vaginal lubrication, and is an essential part of nerve-fiber structure, strength, and resilience.

Reducing cholesterol may also mean lowering sexual activity and becoming nervously irritable and depressed. Fashions in clothing and science always swing sooner or later, and the pendulum in scientific thinking about cholesterol may change in a few years and "fat, happy, sexy, and nervously stable" may be in vogue.

The rap against cholesterol, of course, is that it is found deposited on, and becoming part of, blood vessel walls, reducing the flow of blood through these vessels and putting the person at major health risk. However, if it is partnered by choline and lecithin in appropriate balance, it will not be deposited in arteries but will remain in circulation in the bloodstream—and can improve your sex life and your nervous system health.

Removing as many sources of cholesterol as possible may for a time cause the body to reabsorb and digest stored excess fats, but after between 6 and 12 weeks you'll have lost any excess and will now be seeking treatment for impotence, premature aging, and wrinkling, as well as marriage guidance or sedatives!

Remember, if you cut out all those fats and vegetable oils, you remove 80% to 85% of your sources of the fat-soluble vitamins A, D, and E. Since these are major antioxidants and provide protection for the immune system, cell division, and cell reproduction, you may die not of heart disease, but of various forms of cancer instead. Science doesn't apologize or pay out damages to those who follow its latest breakthrough. Stay with what works. Nature is never wrong, nor has she changed her rules since the world began.

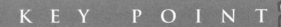

KEY POINT

Isolating a chemical compound from food and using it in very large amounts can cause side effects—like a drug.

Keep your cholesterol balance as nature has arranged for you. Most of the foods and plants that contain cholesterol also contain lecithin and choline. Chinese cuisine is an exquisite blending of this triangle in good natural balance. Most nuts, vegetable oils, and many common plants like fenugreek and dandelion contain the balancing chemistry for cholesterol to remain in circulation—not depositing—as well as for cholesterol itself.

Abnormally high intakes of animal protein (deficient in proteolytic enzymes) and lack of physical exercise are the major offenders in cholesterol deposition problems.

21st Century Nutrition

The 20th century has taught us that the key to good health is nutritional balance, not pharmacologic magic bullets. Let us hope that the 21st century teaches us that common sense and not dollars and cents should be our guide in regulating scientific applications of nutrition and pharmacology knowledge. In that regard, allow me to give you one more glimpse of the future of nutrition. The following article appeared in the *Wisconsin State Journal* on October 5, 1998.

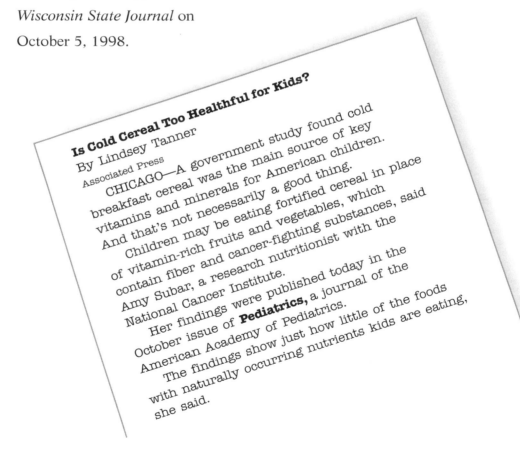

Is Cold Cereal Too Healthful for Kids?
By Lindsey Tanner
Associated Press

CHICAGO—A government study found cold breakfast cereal was the main source of key vitamins and minerals for American children.

And that's not necessarily a good thing.

Children may be eating fortified cereal in place of vitamin-rich fruits and vegetables, which contain fiber and cancer-fighting substances, said Amy Subar, a research nutritionist with the National Cancer Institute.

Her findings were published today in the October issue of **Pediatrics**, a journal of the American Academy of Pediatrics.

The findings show just how little of the foods with naturally occurring nutrients kids are eating, she said.

Kellogg, the world's leading maker of ready-to-eat cereal, sees the study as good news. "It really does show the role cereal can play in delivering important nutrients in kids' diets," spokeswoman Karen Kafer said.

Subar said cereal is not necessarily unhealthy. In fact, fortified cereal appears to be among the more nutritious foods children are eating.

"We just shouldn't kid ourselves that they're necessarily getting all that they need" from cereal, she said.

The study of 4,008 children ages 2 through 18 looked at 20 foods they ate between 1989 and 1991, and cereal was the No. 1 source of vitamin A, iron and folic acid for every age. Folic acid has been shown to reduce the risk of certain types of cancer.

Researchers also found that cereal was the third-highest source of zinc and magnesium.

The researchers did not reach a conclusion on whether the children were getting a nutritionally sound diet, but the findings suggested their diets may be lacking.

For example, sugary drinks were the No. 2 source of carbohydrates, behind bread. And high-fat foods such as cakes and cookies ranked among youngsters' top 10 sources of vitamin E, protein, fiber, calcium and iron.

Gail Frank, a California dietitian and spokeswoman for the American Dietetic Association, said the findings were not surprising but also not terribly worrisome. After all, she said, at least the children weren't skipping breakfast or eating high-fat fast food.

After reading such a report about the attitudes of the professional nutritionists, do you believe the incidence of chronic degenerative diseases will increase or decrease in this country as this generation of schoolchildren matures?

TABLE OF FIGURES

ENZYMES: *The Key to Health*

BIBLIOGRAPHY AND RESOURCE LIST

Adams, Francis, transl. *The Genuine Works of Hippocrates.* Huntington, N.Y.: Robert E. Krieger Publishing Company, 1972.

Arey, Leslie Brainerd. *Developmental Anatomy: A Textbook and Laboratory Manual of Embryology.* 6th ed. Philadelphia: W. B. Saunders Company, 1959.

Berdanier, Carolyn D. *CRC Desk Reference for Nutrition.* New York: CRC Press, 1998.

Bieler, Henry G. *Food is Your Best Medicine.* New York: Random House, 1965.

Bloom, William and Don W. Fawcett. *A Textbook of Histology.* 8th ed. Philadelphia: W. B. Saunders Company, 1962.

Carter, Richard. *Your Food and Your Health.* New York: Harper & Row, 1964.

Dubos, Rene. *The Mirage of Health.* New York: Harper & Brothers Publishers, 1959.

Duke, James A. *Handbook of Medicinal Herbs.* New York: CRC Press, 1985.

Dunne, Lavon J. *Nutrition Almanac.* 3d ed. New York: McGraw-Hill, 1990.

Gibson, Glenn R. and George T. Macfarlane, eds. *Human Colonic Bacteria: Role in Nutrition, Physiology, and Pathology.* Ann Arbor: CRC Press, 1995.

Greenblatt, Robert B. *Office Endocrinology.* 3d ed. Springfield, Ill.: Charles C. Thomas, 1947.

Guyton, Arthur C. *Textbook of Medical Physiology.* 7th ed. Philadelphia: W. B. Saunders, 1986.

Hasselberger, Francis X. *Uses of Enzymes and Immobilized Enzymes.* Chicago: Nelson-Hall, Inc., 1978.

Holt, Peter R. and Robert M. Russell. *Chronic Gastritis and Hypochlorhydria in the Elderly.* Ann Arbor: CRC Press, 1993.

Howell, Edward. *Enzyme Nutrition: The Food Enzyme Concept.* Wayne, N.J.: Avery Publishing Group Inc., 1985.

—. *Food Enzymes for Health & Longevity*. Woodstock Valley, Conn.: Omangod Press, 1980.

Immerman, Alan. "Evidence for Intestinal Toxemia: An Inescapable Clinical Phenomenon." The American Chiropractic Association Insert, *Journal of Chiropractic,* 13 (1979)

Kleiner, Israel S. *Biochemistry.* 6th ed. St. Louis: C. V. Mosby Company, 1962.

Lehninger, Albert L. *Principles of Biochemistry.* New York: Worth Publishers, 1982.

Loomis, Howard F., Jr., D.C., F.I.A.C.A., *The Nutritional Value and Clinical Application of Food Enzymes.* N.p., 1986.

Loomis, Justin R. *Elements of the Anatomy, Physiology, and Hygiene of the Human System.* New York: Sheldon, Blakeman & Co., 1856.

National Research Council. *Recommended Dietary Allowances.* 10th ed. Washington, D.C.: National Academy Press, 1989.

Pottenger, Francis M., Jr. *Pottenger's Cats: A Study in Nutrition.* 2d ed. N.p., 1995.

Price, Weston A. *Nutrition and Physical Degeneration.* 6th ed. New Canaan, Conn.: Keats Publishing, Inc., 1998.

Rideal, Samuel. *Disinfection and the Preservation of Food.* New York: John Wiley and Sons, 1903.

Rose, Steven. *The Chemistry of Life.* 2d ed. New York: Penguin Books, 1979.

Schultz, H. W., ed. *Food Enzymes.* Vol 1. Symposium on Food. Westport, Conn.: The Avi Publishing Company, Inc., 1960.

Selye, Hans. *In Vivo: The Case for Supramolecular Biology.* New York: Liveright Publishing Corporation, 1967.

—. *The Stress of Life.* New York: McGraw-Hill, 1956.

Shils, Maurice E., James A. Olson, and Moshe Shike, eds. *Modern Nutrition in Health and Disease.* 2 vols. 8th ed. Philadelphia: Lea & Febiger, 1994.

Spiro, Howard M. *Clinical Gastroenterology.* 3d ed. New York: Macmillan Publishing Co., Inc., 1983.

Tilden, John H. *Impaired Health: Its Cause and Cure.* 2 vols. 3d ed. Mokelumne Hill, Calif.: Health Research, 1959.

———. *Toxemia Explained.* Rev. ed. Mokelumne Hill, Calif.: Health Research, 1960.

U.S. Department of Health and Human Services. *The Surgeon General's Report on Nutrition and Health.* Washington, D.C.: GPO, 1989.

Vander, Arthur J., James H. Sherman, and Dorothy S. Luciano. *Human Physiology: The Mechanisms of Body Function.* 4th ed. New York: McGraw-Hill, 1985.

Virchow, Rudolf. *Cellular Pathology.* 1858. Reprint, Birmingham: Gryphon Editions, Ltd., The Classics of Medicine Library, 1978.

Wapnir, Raul A. *Protein Nutrition and Mineral Absorption.* Boca Raton: CRC Press, 1990.

Williams, Roger. *You Are Extraordinary.* New York: Pyramid Books, 1974.

Williams, Sue Rodwell. *Nutrition and Diet Therapy.* 5th ed. St. Louis: Times Mirror/Mosby College Publishing, 1985.

Wolf, Max and Karl Ransberger. *Enzyme Therapy.* New York: Vantage Press, Inc., 1972.

HERBAL BIBLIOGRAPHY

Bisset, Norman Grainger, ed. and trans. 2d German ed. *Herbal Drugs and Phytopharmaceuticals.* Stuttgart: Medpharm Scientific Publishers; Boca Raton: CRC Press, Inc., 1994.

Blumenthal, Mark. *The Complete German Commission E Monographs: Therapeutic Guide to Herbal Medicines.* Austin, Tex.: American Botanical Council; Boston, Mass.: Integrative Medicine Communications, 1998.

Duke, James A., Ph.D. *Handbook of Medicinal Herbs.* Boca Raton: CRC Press, Inc., 1985.

Huang, Kee Chang. *The Pharmacology of Chinese Herbs.* 2d ed. Boca Raton: CRC Press, Inc., 1999.

PDR for Herbal Medicines. 1st ed. Montbale, N.J.: Medical Economics Company, Inc., 1998.

GLOSSARY

The following is not intended to replace a medical dictionary, but rather to assist the reader in understanding the terms as they are used within the context of this book.

A

acidosis
A condition characterized by the accumulation of acid or loss of alkalinity. Also see *alkalosis.*

active ingredient
A term used to identify the chemical compound in a food or herb which can be isolated away from its source and used, as a drug, to achieve a desired response. The same ingredient when consumed in the food or herb containing the synergistic compounds needed for its assimilation does not produce the same effect.

active transport mechanism
The energy-consuming process the body uses to move a nutrient in the blood into an organ where its concentration is much higher than in the blood. For example, the concentration of iodine within the thyroid gland may be as much as 25 to 300 times higher within the gland than it is in the blood.

acupuncture
This is a Western term used to indicate the use of needles to puncture the skin to gain a therapeutic response. In China, this method is only part of an ancient healing method that uses herbs (nutrition), spinal manipulation (chiropractic), and needles (physical therapy) to direct the body's own innate healing energies.

adsorb
This term means to cling or attach to the outside of a cell as opposed to being taken into the cell through absorption.

aggregated lymphatic follicles
Small sacs of lymphatic tissue filled with white blood cells (phagocytes) spaced throughout the body like barracks or forts to house white blood cells that attack foreign invaders. Also see *Peyer's patches* and *Kupffer cells.*

alkaloids

Nitrogen-containing chemical compounds found in plants. These are usually the active ingredients that pharmaceutical companies isolate from plants and concentrate to achieve specific effects in treatment of disease conditions.

alkalosis

A condition characterized by the accumulation of alkalinity or loss of acidity. Also see *acidosis.*

alveoli

Small spaces or cavities in the lungs containing air.

amylase

A carbohydrate-digesting enzyme.

anoxia

A condition describing a lack of oxygen.

antibiotic

Literally means anti-life. Generally refers to chemical compounds developed to inhibit bacterial growth.

antibody

A protective, immune protein designed to react with foreign agents that can enter the body.

antigen

Any substance that produces an immune reaction in the body. Also see *antibody.*

apoenzyme

The protein portion of an enzyme, as compared to the coenzyme or prosthetic portion.

appetite

The desire or longing to satisfy any physical or mental need. In this book, it refers to the desire for food. Also see *satiety.*

-ase

The ending that refers to an enzyme, as in protease (protein + -ase).

aspergillus

A species of fungi uniquely suited for growing enzymes. The final enzyme product does not contain any aspergillus when it is placed into dietary supplements.

autointoxication

The process of self-poisoning. In this book it refers to the process of toxins (poisons) forming in the intestine from the fermentation of inadequately digested foods. These toxins cause an inflammatory response in the wall of the intestine. They also are absorbed into the blood where they cause many harmful effects within the body.

autolysis

The process in which the body uses the enzymes contained within a dead or degenerated cell to digest the cell itself.

autophagia

The process of the body to maintain normal homeostasis (function and nutrition) by digesting its own cells. For example, as in severe deficiencies or starvation.

B

bacteria

One-celled microorganisms that reproduce via asexual (no partner) cell division and possess a cell membranc for uniformity of size and shape. They are parasitic and pathogenic (disease-producing). Also see *viruses.*

basophils

Specialized white blood cells that contain heparin, which is used by the body to thin the blood when needed. Also see *granulocytes.*

bruxism

The clenching or grinding of teeth, usually during sleep.

C

calorimetry

The measurement of the amount of heat given off during a chemical reaction or group of reactions.

carbohydrate

A substance found in food that is composed of simple sugars (from fruits, dairy products, grains and flour, white sugar, honey, molasses, and maple syrup to name a few), complex starches (vegetables), and fiber.

catalase

An oxidoreductase enzyme made in every cell of the body. The enzyme is essential to destroy the hydrogen peroxide (H_2O_2) formed when glucose is converted into energy. H_2O_2 is a poison to the body and must be destroyed. Catalase changes the poisonous hydrogen peroxide into water and oxygen, two substances the body can use.

catalyst

A substance that must be present for a chemical reaction to take place, but itself is not changed during the reaction. In this book, it refers to enzymes. While they are not changed during the digestive process, they nevertheless eventually wear out and must be replaced.

cellulase

An enzyme that digests the fiber found in vegetables and some fruits.

cellulose

A large carbohydrate molecule (polysaccharide) that forms vegetable fiber. The body does not make an enzyme to digest cellulose, but cellulase is contained within the raw vegetables. When the fibrous foods are chewed, the cellulase will digest the fiber in the upper part of the stomach.

chelated

An expensive process combining a mineral with another substance, such as protein, in hopes of moving it out of the intestine and into the blood. Minerals contained in food are already chelated and only require adequate digestion to be utilized by the body.

cholecystokinin

A hormone formed in the wall of the small intestine when stimulated by the acidic contents of the stomach. This hormone stimulates the contraction of the gallbladder and the release of bile to emulsify (degrease) fats. Unless this is done, enzymes cannot penetrate the fats to digest the food.

cholesterol

A soft, waxy substance made in the liver and found in certain foods. The body needs cholesterol for normal brain function and hormone production.

chyme

The semifluid, partly digested food that passes from the stomach into the first part of the small intestine (duodenum).

circulating immune complexes

A relatively new term used to describe the combination of the body's defensive reaction (antibody) with a foreign invader (antigen).

coenzyme

A substance that is necessary to, or enhances the action of, an enzyme. Usually a vitamin or mineral.

colloidal suspension

Aggregates of atoms or molecules in a finely divided state and dispersed in a gas, liquid, or solid. The particles then resist settling out of the substance they are contained in. Minerals found in food and juices are in colloidal suspension and do not require special preparation.

constipation

A term commonly used to indicate a hard stool and/or difficult or painful bowel movements. It should be used to indicate less than one bowel movement per day.

D

dermatoses

Dermatoses is the plural form of dermatosis and is a general term used to indicate an abnormal condition of the skin.

diarrhea

A term commonly used to indicate a loose or liquid stool. It should be used to indicate frequent and soft bowel movements.

digestion

The process by which a substance (food) is divided into two or more substances by enzymes. Water is used in the process. For example, an enzyme breaks a food into two parts and adds a molecule of hydrogen (H+) to one broken end, and a hydroxyl molecule (OH-) to the other broken end, thus effectively using a molecule of water to form two new, smaller compounds. This is how food is digested. Also see *hydrolysis.*

digestive leukocytosis

The elevation of the number of white blood cells in the circulating blood (leukocytosis) that occurs within 30 minutes after eating enzyme-deficient food.

disease

Illness or sickness that indicates an inability of the body to maintain normal function(s).

drug

Any substance other than food used for the diagnosis, alleviation, treatment, or cure of disease.

E

eclampsia

Used to describe convulsions that are not caused by a condition of the brain. The term is used in association with a pregnancy in which there is high blood pressure and the presence of protein in the urine. When it occurs, this condition usually happens during the 20th week of pregnancy.

ectoderm

The outer layer of the three layers of an embryo. This layer of the embryo later forms the skin, nerves, and central nervous system of the developing fetus.

eczema

A general term used to indicate an inflammatory condition of the skin, often characterized by itching, burning, or scaling.

electrolyte

Any compound that when it is dissolved in water can carry an electrical current. A compound that becomes ionized in solution, such as sodium hydroxide (Na+ OH_3-).

endoderm

The inside layer of the three layers of an embryo. This layer of the embryo later forms the digestive tract and other organs in the developing fetus.

enzyme

Organic catalysts that are composed of protein combined with either a mineral, vitamin, or part of a vitamin. They are responsible for performing chemical changes in the body without themselves being changed in the process.

Enzyme Replacement Nutrition

The process of supplementing naturally occurring food enzymes into the diet to replace those that are removed or deficient in the modern diet.

eosinophils

Specialized white blood cells used by the body to react against allergens and parasites. Also see *granulocytes.*

epithelial tissue

The outer layer of skin covering the outside surfaces of the body. Also lining the digestive, urinary, and respiratory tracts. Substances passing through those structures are considered to be outside the body.

extracellular fluid (ECF) compartment

Body fluids that are outside the cells, including the blood. Comprises approximately one-third of all body fluids. It is this fluid that must be maintained within relatively narrow limits of temperature, pH, volume, and concentration of the dissolved substances it contains. The dissolved substances refers to those substances we test blood for, such as cholesterol, glucose, and triglycerides.

F

fats or lipids

A semisolid substance found in food and composed of glycerol (a sweet-tasting oil) and fatty acids (any acid derived from fat).

fibrin

Fibrin is an elastic, protein material involved with the clotting of blood and formation of scabs. Also see *plasmin* and *thrombin.*

fibromyalgia

A painful condition affecting the muscles and connective tissues of the body. It is not contagious, is not progressive, and does not cause death. Once thought to be psychosomatic, medicine has been unable to find its cause.

flora

Refers to microorganisms normally found on the external surfaces of humans/animals. These include the surfaces of the gastrointestinal tract—the lumen of which is considered to be outside of the body—and the respiratory and urinary tracts.

fluid transport

The blood is a fluid that is used to transport nutrients to the cells and to transport cellular waste away from the cells.

food

Any substance usually eaten by humans for nourishment. Raw food contains protein, carbohydrates, fats or lipids, vitamins, minerals, and enzymes.

Food Enzyme Concept

The theoretical work of Dr. Edward Howell stressing the inclusion of enzymes from plant foods in the diet and their importance in digesting food in the stomach before stomach acid can be produced, thus relieving digestive stress and pancreatic hypertrophy.

fructose

A simple sugar (monosaccharide) found in vegetables and fruits. Fructose combined with a molecule of glucose forms the disaccharide sucrose. When liberated by the action of sucrase, the fructose is sent to the liver where it is converted to glucose for energy.

G

galactose

A simple sugar (monosaccharide) found in dairy products. Galactose combined with a molecule of glucose forms lactose (milk sugar). When liberated by the action of lactase, the galactose is sent to the liver where it is converted to glucose for energy.

gastrointestinal tract
Relating to the stomach and intestines.

genitourinary
A general term used to describe tissues or conditions affecting the urinary and reproductive organs of the body.

globulins
A family of special proteins formed in the body, especially the liver, that contain the body's antibodies and are therefore the foundation of our immune system.

glucose
A monosaccharide, it is the desired end product of all carbohydrate digestion. All carbohydrates and some proteins eventually become glucose as a result of normal digestion. Stored fat can be converted to glucose when needed. Glucose is the sugar needed by the cells for energy formation.

glycerol
A sweet-tasting oil contained in fats that is used to transport fatty acids.

glycosides
A chemical compound derived from sugar or sugar-related.

goiter
A chronic enlargement of the thyroid gland, not due to tumor growth. It is associated with the body's attempt to make more thyroid tissue in a vain effort to make more thyroid hormone.

granules
A term used to describe a grain-like substance (such as a grain of sand). For example, the granules seen in some white blood cells are actually small discrete masses of enzymes.

granulocytes
Specialized white blood cells that contain granules which absorb dyes that can be used for identifying them under a microscope. Also see *neutrophils, eosinophils,* and *basophils.*

H

halitosis
The condition of bad breath.

hepatotoxicity
The liver is a major organ of detoxification in the body. The term hepatotoxicity implies that the organ is overwhelmed with toxicity and unable to perform properly.

herbology
The time-honored study of the effects of herbs (plants) on the human body.

heterogeneous
Containing parts having various and dissimilar characteristics or properties, as opposed to homogeneous where the parts are similar. Also see *homogeneous.*

homeopathy
A system of therapy that uses minute amounts of a substance that can produce certain effects within the body to treat diseases or conditions that have the identical symptoms.

homogeneous
Having a uniform consistency, each part looks like the other parts. Also see *heterogeneous.*

hydrochloric acid
A very strong acid formed by the combination of H+ (hydrogen) and Cl- (chloride) ions to form stomach acid.

hydrolysis
The process by which a substance (food) is divided into two or more substances by enzyme. Water is used in the process. For example, an enzyme breaks a food into two parts and adds a molecule of hydrogen (H+) to one broken end, and a hydroxyl molecule (OH-) to the other broken end, thus effectively using a molecule of water to form two new and smaller compounds. This is how food is digested. Also see *digestion.*

hyperpermeability

To be permeable means to allow passage of substances through membranes of the body. Hyperpermeability means that the membrane allows a greater amount of those substances to cross, even allowing substances to cross that normally could not.

hyperphagia

Overeating. Compare to aphagia which indicates a failure to eat, or dysphagia which means it hurts to swallow.

hypertrophy

Describes enlargement of a body organ, but not involving tumor growth. Usually suggests enlargement of an organ, such as a thyroid or pancreas, in a vain effort to provide increased secretion to maintain health.

hypothalamus

An organ in the brain that is primarily involved with regulating the autonomic nervous system and the hormonal system of the body, the two major control mechanisms by which the body maintains its normal functions. It also influences our sense of taste and smell.

hypothyroidism

A condition characterized by insufficient thyroid hormone, weight gain, lack of energy, and need for unusual amounts of sleep.

I

iatrogenic

Doctor-induced. An unfavorable body response to medical or drug treatment caused by the treatment itself.

ileum

The third and last part of the small intestine. About 12 feet long, it extends from the jejunum to the ileocecal valve that separates the small and large intestine.

immunopathies

Pathological (disease) conditions involving the immune system.

immunoreactive

Denotes a reaction of the immune system, especially between a foreign invader (antigen) and the body's defensive reaction (antibody). Also see *circulating immune complexes*.

indican

A substance found in urine that indicates the rotting or decaying (putrefaction) of protein in the intestine.

indicanuria

The presence of indican in urine.

inflammation

A defensive immune reaction on the part of the body in response to some irritant, infestation, or infection. Usually implies the presence of redness, fever, swelling, or pain.

inhibitors

A chemical that prevents a normal reaction from occurring such as the enzyme inhibitors found within seeds and nuts. These inhibitors prevent the enzymes within the seeds and nuts from working thus preventing or inhibiting digestion. They can be removed by cooking or roasting, which also destroys the enzymes themselves. See page 75 for competitive and noncompetitive inhibitors.

internal environment (or intracellular fluid compartment)

The fluid contained inside the cells of the body comprises about two-thirds of all body fluids.

interstitial fluid

Fluid contained within the spaces between the cells, but not within a body cavity. This fluid would be considered extracellular.

intoxication

Implies poisoning by any substance, such as alcohol.

isomerization

The process by which a mirror image of a chemical compound is formed. Identical chemical structures, but reversed as in a mirror, and having different physical and chemical properties.

J

jejunum

The second portion of the small intestine. After food leaves the duodenum, it enters the jejunum which secretes the disaccharidases—lactase, maltase, and sucrase—that digest simple sugars.

K

kinin

One of many substances released in the blood by the breakdown of protein as part of some pathological process. Usually affecting blood flow, some kinins increase blood to a part of the body, while others decrease blood flow.

Kupffer cells

A specialized defensive immune tissue in the liver, identical to the tonsils, that contain concentrations of white blood cells to counteract foreign materials that have entered the blood. Also see *Peyer's patches*.

L

lactase

Lactase is a simple sugar-digesting enzyme (disaccharidase) that digests milk sugar (lactose). The gas, bloating, and diarrhea that occurs when the small intestine cannot produce enough lactase to digest the lactose contained in the excessive milk or ice cream eaten is called lactose intolerance.

lactobacillus

Lactic acid-forming, anaerobic (do not require oxygen), nonmoving bacteria, which are involved in a complex number of nutritional processes. They are generally not pathogenic (do not cause disease) and are considered a part of the normal flora of the mouth, intestine, and vagina. The term flora refers to the microorganisms normally found in these tissues.

lactose

The sugar found in dairy products, especially milk and ice cream.

leukocytes

Used interchangeably with the term white blood cells.

leukocytosis

A transient increase in the number of white blood cells circulating in the blood occurring in response to such events as hemorrhage, injury, inflammation, fever, or entry of foreign material into the body (infection or infestation, for example). As opposed to leukemia, which is a progressive malignant disease of the blood-forming organs.

lipase

A fat-digesting enzyme.

lumen

The space in the interior of a tube, such as an artery or the intestine, through which the blood or food is moved.

lymphocytes

Specialized white blood cells, which do not contain granules, used by the body to fight infection and respond to irritated or injured tissue. Also see *phagocytes*.

lysosomes

Pockets of enzymes enclosed within their own membrane inside the cytoplasm of a cell. They contain hydrolytic (digestive) enzymes that digest foreign matter that has been engulfed by the cell.

M

macrophagocytes

Very large white blood cells that use enzymes to literally "eat" foreign particles in the body. Also see *phagocytes*.

magic bullet

A term in common usage to describe the search for chemical compounds that can alleviate or cover up symptoms of disease without determining and removing the cause of the disease.

malassimilation

Incomplete or faulty assimilation of food. Usually implies the inability to move nutrients across the gut wall into the blood.

maltase

Maltase is a simple sugar–digesting enzyme (disaccharidase) that digests the sugar found in grains (maltose), such as wheat. The gas, bloating, and diarrhea that occurs when the small intestine cannot produce enough maltase to digest the maltose contained in grains or flour is called gluten intolerance.

maltose

The sugar found in grains and flours.

mesoderm

The middle layer of the three layers of an embryo. This layer in the embryo forms the muscles, tendons, ligaments, and connective tissues of the developing fetus.

microglia

Small cells found in the nervous system that may have phagocytic properties.

microthrombi

Small blood clots.

microvilli

Hair-like projections from the wall of the small intestine. Simple sugar–digesting enzymes originate from here.

mineral

Inorganic materials found in the earth's crust. They cannot be made in the human body. They are found in small quantities in food and are essential for human function and health.

mitosis

A normal process of a cell dividing into two identical daughter cells.

monocytes

Specialized white blood cells, which do not contain granules, used by the body to fight infection and respond to irritated or injured tissues. Also see *phagocytes.*

myofascitis

An inflammatory condition of the fascia (tissues that lie under the skin and cover and separate muscles and groups of muscles). Also see *fibromyalgia.*

N

neutrophils

Specialized white blood cells, which contain granules, used by the body to fight infection and respond to irritated or injured tissues. Also see *granulocytes.*

NSAIDs (nonsteroidal anti-inflammatory drugs)

Pain relievers such as ibuprofen and naprosyn.

nutrition

The process of ingesting food and converting it into energy or using it to repair body tissue.

nutritional objectivity

The process of selectively choosing dietary substances based on objective individual clinical evidence as opposed to a hit-or-miss process used to relieve symptoms.

O

opsonins

A substance that encourages or promotes phagocytosis. The body actually covers bacteria and other foreign particles with opsonins to facilitate their recognition by the white blood cells (phagocytes).

organelles

A general term used to describe the various structures found within a cell.

-ose

This ending indicates a carbohydrate, for example, fructose, glucose, sucrose, or lactose.

oxidation

The process of combining with oxygen. The oxidized substance gives up electrons. Also see *reduction.*

P

palpation

The art of examining the body by using the hands to feel muscle contractions.

pancreatic hypertrophy

Describes enlargement of the pancreas (but not involving tumor growth) in a vain effort to provide increased enzyme secretion. Caused by the inability of the pancreas to provide all the enzymes needed to digest the food eaten. This condition does not occur suddenly but is acquired over years of dietary indiscretions.

parasites

An organism that lives on or in another organism and draws its nutrition from its host, such as intestinal worms and various fungi/yeast such as *Candida albicans*.

parasympathetic

This is one of two divisions of the autonomic nervous system. It controls functions that assist the body during rest, recuperation, and reproductive situations. Also see *sympathetic* and Table 5 on page 53.

paresthesia

Any abnormal sensation such as burning, itching, or tingling.

parotid glands

Located near the ear, these salivary glands secrete the carbohydrate-digesting enzyme amylase. Also see *submandibular* and *sublingual glands*.

pathogen

Any microorganism, such as a virus, or other substance that causes a disease.

peristalsis

The rhythmic muscular contractions of the walls of the gastrointestinal tract by which the body moves food through the digestive system.

peritoneum

The serous (watery) membrane that lines the abdominal cavity and covers most of the organs in the abdomen.

Peyer's patches

A specialized defensive immune tissue in the mucosal lining of the intestine, identical to the tonsils, that contain concentrations of white blood cells to prevent foreign materials from passing into the blood. Also see *Kupffer cells*.

pH

A scale used to indicate the acidity or alkalinity of a substance. The term is an acronym for potential of hydrogen. The pH is graded on a scale of 1 to 14. Acid substances have a number below 7, and alkaline substances have a number above 7. Pure water has a pH of 7.0. Blood has a pH generally around 7.4 and is considered slightly alkaline.

phagocytes

White blood cells that use enzymes to literally eat foreign particles in the body. Also see *macrophagocytes.*

phagocytosis

The process of white blood cells engulfing foreign materials in the blood and using their enzymes to literally eat (digest) the invader. Also see *pinocytosis.*

pharmacology

The study of drugs and their effects.

phytopharmacy

A relatively new term used to designate nutrients contained in plants that can be removed and given in concentrated doses.

pinocytosis

The process by which a stationary cell (not a moving white blood cell capable of independent movement) ingests a foreign particle that has landed on it and uses its own enzymes to destroy (digest) it. Also see *phagocytosis.*

plasmin

A protein-digesting enzyme that hydrolyzes (digests) fibrin. Also see *fibrin* and *thrombin.*

pleura

A serous (watery) membrane that lines and protects the lungs.

postcentral gyrus

Refers to a specific area of the brain.

predigestion

The process of enzymes digesting food in the stomach before stomach acid can be produced.

ENZYMES: *The Key to Health*

prostaglandins and prostaglandin inhibitors
A class of active substances found in many tissues that produce a wide variety of specific body reactions when stimulated. Prostaglandin inhibitors thus block those reactions from occurring.

protease
A protein-digesting enzyme.

protein
A substance found in food that contains nitrogen and is composed of amino acids. Examples of protein-rich foods include meats, cheese, and eggs.

proteolytic
A term that describes the digestion of proteins by protease.

psoriasis
Describes an eruption of round and red lesions covered with silvery scales on the body. Usually found on the elbows, knees, scalp, chest, and abdomen.

putrefaction
The process of decaying or rotting. It implies the breakdown of organic material (food) by bacteria action in the colon with the formation of chemicals (such as indican) and gases.

pyloric valve
The valve that opens from the stomach into the small intestine (duodenum).

Q

R

reduction
The process of gaining one or more electrons. This occurs when hydrogen (H+) is added to the double bond of an organic molecule. The hydrogen is reduced (gains an electron), and the organic molecule is oxidized (loses an electron). Also see *oxidation*.

respiration
The process of breathing by which the body takes on oxygen and gives up carbon dioxide.

reticuloendothelial system

That part of the immune system composed of phagocytes, as well as the cells lining sinusoids of the spleen, liver, lymph nodes, and bone marrow. A front line of defense to prevent foreign matter from entering the body through the digestive tract, and cleansing the blood when foreign particles do breach the system.

rhinitis

Inflammation of the nasal passages.

S

saponins

Glycosides (sugar-derived) plant substances that foam in water and are derived by the action of alkalis of fat, thus forming soaps.

satiety

The recognition by the hypothalamus gland that the desire or longing for food has been satisfied. The sensation of feeling full. Also see *appetite.*

secretin

A hormone formed in the wall of the small intestine (duodenum) when stimulated by the acidic chyme from the stomach. This hormone stimulates the release of enzymes and alkalinity by the pancreas to continue the digestive process.

sinusoidal cells

These are cells, supplied by small thin-walled blood vessels, found in the sinuses or spaces of the body.

stress

Any mechanical, chemical, or emotional stimulus that exhausts the normal processes of the human body.

sublingual glands

Located under the tongue, these salivary glands secrete the fat-digesting enzyme lipase. Also see *submandibular* and *parotid glands.*

submandibular glands

Located under the jaw, these salivary glands secrete the protein-digesting enzyme protease. Also see *submandibular* and *sublingual glands.*

sucrase

Sucrase is a simple sugar–digesting enzyme (disaccharidase) that digests the sucrose found in foods such as white sugar, molasses, honey, maple syrup, white flour. Constipation and irritability can occur when the small intestine cannot produce enough sucrase to digest the sucrose contained in the average diet.

sucrose

Normally occurs in beets and cane sugar. It is composed of a molecule of fructose and a molecule of glucose and requires the enzyme of sucrase for digestion.

sympathetic

One of two divisions of the autonomic nervous system. It controls functions that assist the body during "fright, fight, or flight" situations. Also see *parasympathetic* and Table 5 on page 53.

symptom

Any sensation experienced by a person that is indicative of a departure from normal function or structure.

symptomatology

The study of the symptoms involved in a patient's condition or illness.

T

tachycardia

A rapid beating of the heart. Implies a pulse rate over 100 beats per minute.

tannins

Complex constituents of plants that have astringent or tightening effects on tissues.

thrombin

A protein-digesting enzyme that converts fibrinogen into fibrin to form a scab in blood that has left the body. Also see *fibrin* and *plasmin*.

U

urinalysis

A collection of tests, both chemical and physical, used to examine the contents of the urine.

V

vesicle

A small, round elevation of the skin that contains fluid such as a blister.

viruses

A group of very small one-celled microorganisms that do not have a cell membrane as bacteria do. Viruses are so small they can pass through filters that bacteria cannot. Viruses are incapable of growth and reproduction without becoming part of a living cell. Also see *bacteria*.

vitamin

Organic substances that cannot be made in the human body. They are found in small quantities in food and are essential for normal human function and health.

W

X

Y

Z

INDEX

postcentral gyrus, 43

prostaglandin inhibitors, 115

protease, xxv, 99

putrefaction, 110

pyloric valve, 97

pylorus, 97

Q

R

reduction, 76

respiration, 5, 20, 53

reticuloendothelial system, 127, 128, 129

S

salicylic acid, 64, 65, 66, 67

saponins, 154

secretin, 97, 98

Selye, Hans, 20, 21, 22, 25, 144

sinusoidal cells, 113

stomach acid, 68, 80, 92, 95, 96, 98, 116, 144

stress, 9, 15, 20, 21, 22, 23, 24, 27, 39, 50, 111, 147

Stress of Life, The 20, 22

sublingual glands, 100

submandibular glands, 91, 101

sucrose, 81, 82, 83, 99, 102, 103

Surgeon General, 10, 29, 30, 137

sympathetic, 46, 52, 54

symptom, 7, 12, 17, 18, 23, 52, 104, 138

symptomatology, 12, 13, 14

T

tannins, 153

testis, 48

thrombin, 122

thyroid, 48, 89

U

U.S. Recommended Daily Allowances, 61, 64

urinalysis, xxiii, xxvi, xxvii, 27, 128

V

vesicle, 126

W

Williams, Roger J., 15

X

Y

Z

ENZYMES: *The Key to Health*